Julia Perkins Pratt Ballard

Among the Moths and Butterflies

Julia Perkins Pratt Ballard

Among the Moths and Butterflies

ISBN/EAN: 9783744758529

Printed in Europe, USA, Canada, Australia, Japan

Cover: Foto ©berggeist007 / pixelio.de

More available books at **www.hansebooks.com**

AMONG THE MOTHS

AND

BUTTERFLIES

A REVISED AND ENLARGED EDITION OF
"INSECT LIVES; OR, BORN IN PRISON"

BY

JULIA P. BALLARD

———

G. P. PUTNAM'S SONS

NEW YORK   LONDON
27 WEST TWENTY-THIRD ST.   27 KING WILLIAM ST., STRAND

The Knickerbocker Press

1890

The Knickerbocker Press, New York
Electrotyped, Printed, and Bound by
G. P. Putnam's Sons

TO

MY SON

HARLAN HOGUE BALLARD

CONTENTS.

	PAGE
PREFACE TO "INSECT LIVES"	. xvii
PREFACE TO REVISED EDITION	. xxi
INTRODUCTORY	xxiii

CHAPTER I.
BORN IN PRISON	1

CHAPTER II.
THE GREEN HOUSE WITH GOLD NAILS .	4

CHAPTER III.
TWO FRONT DOORS, AND WHAT WAS BEHIND THEM .	17

CHAPTER IV.
THE EARLY BUTTERFLY .	. 25

CHAPTER V.
THROUGH A GLASS CLEARLY .	. 30

CHAPTER VI.
HOW I CAUGHT A BEAR .	. 44

CHAPTER VII.
CRUMPLE-WING . . .	49

CHAPTER VIII.
UNDER THE CAPE . .	. 53

CHAPTER IX.
THE ARCTIAN AND ICHNEUMON	56

CHAPTER X.

PAGE

THE WHITE ERMINE MOTH . . 58

CHAPTER XI.

A HUNDRED TO ONE 59

CHAPTER XII.

THE UNFINISHED LIFE OF QUAKER GRAY . 68

CHAPTER XIII.

AN EARLY CECROPIAN 70

CHAPTER XIV.

THE ROSY DRYOCAMPA 79

CHAPTER XV.

THE SATURNIA IO . . 86

CHAPTER XVI.

SILVER GRAY 95

CHAPTER XVII.

THE CERATOMIA QUADRICORNIS . IOI

CHAPTER XVIII.

PHILAMPELUS ACHEMON . . . 105

CHAPTER XIX.

THE FOX-FACED MOTH [ADONETA SPINULOIDES] . . 110

CHAPTER XX.

LIFE IN A BASKET 115

CHAPTER XXI.

A BLACKBERRY LOOPER . . 119

CHAPTER XXII.

THE DRYOCAMPA IMPERIALIS . . . 122

CHAPTER XXIII.

A BARREL FULL OF LUNAS 128

CONTENTS.

CHAPTER XXIV.

PAGE

THE FEBRUARY BUTTERFLY [PAPILIO CRESPHONTES] . 135

CHAPTER XXV.

A THOUSAND TO ONE . . . 144

CHAPTER XXVI.

THE COMPLAINT OF THE CHRYSALIS . . . 149

CHAPTER XXVII.

THE TUSSOCK MOTH . . . 151

CHAPTER XXVIII.

WINGED AND WINGLESS . . 156

CHAPTER XXIX.

A RACE FOR LIFE 163

CHAPTER XXX.

THE BULRUSH CATERPILLAR . . . 168

CHAPTER XXXI.

A BEADED CATERPILLAR . . . 174

CHAPTER XXXII.

ATTACUS CYNTHIA 177

CHAPTER XXXIII.

THE TURNUS BUTTERFLY . . . 181

CHAPTER XXXIV.

THE BEECH-NUT BOX [LIMACODES SCAPHA] . . 187

CHAPTER XXXV.

THE " MONKEY-FACED " MOTH—HAG MOTH [PHOBETRON
PITHECIUM] 195

CHAPTER XXXVI.

THE SMARTWEED CATERPILLAR 201

CHAPTER XXXVII.

PAGE

THE GREAT LEOPARD MOTH .   . 206

CHAPTER XXXVIII.

A BUTTERFLY CHASE   .   . 212

CHAPTER XXXIX.

TWO SIDES TO A SHIELD, THE WHITE-LINED MORNING
     SPHINX [DEILEPHILA LINEATA]   .   .   . 217

CHAPTER XL.

THE " DECEPTIVE MOTH "   . 223

CHAPTER XLI.

THE ROYAL WALNUT MOTH .   .   . 227

ILLUSTRATIONS.

FIGURE		PAGE
	AMONG THE MOTHS AND BUTTERFLIES . *Frontispiece.*	
	GALL-NUTS ("OAK APPLES") .	XVII
	" BEECHNUT-BOX " .	XX
1.	GROOVED EGG .	XXXIII
2, 3.	CABBAGE BUTTERFLY, CATERPILLAR, AND CHRYSALIS	2
4, 5.	DANAIS CATERPILLAR AND EGG (MAGNIFIED)	5
6.	DANAIS CATERPILLAR SUSPENDED FOR CHRYSALIS	6
7, 8, 9.	DIFFERENT VIEWS OF DANAIS CHRYSALIS	7
10.	DANAIS ARCHIPPUS BUTTERFLY	10
11.	DANAIS CHRYSALIDS, BUTTERFLY, AND CATERPILLAR	16
12.	CATERPILLAR OF PAPILIO ASTERIAS	17
13.	PAPILIO ASTERIAS	19
14.	CHRYSALIDS OF PAPILIO ASTERIAS .	20
15.	SUSPENDED ASTERIAS CHRYSALIS	21
16.	ICHNEUMON FLIES	22
17.	THE EARLY BUTTERFLY : VANESSA ANTIOPA .	25
18.	VANESSA ANTIOPA CATERPILLAR	26
19.	CHRYSALIS OF VANESSA ANTIOPA .	29
20.	POLYPHEMUS COCOON	30
21, 22.	BACK AND FRONT VIEW OF POLYPHEMUS CHRYSALIS	31
23.	POLYPHEMUS ANTENNA	31
24.	POLYPHEMUS MOTH	33
25.	POLYPHEMUS CATERPILLAR	35

FIGURE		PAGE
26.	Yellow-Bear Caterpillar	45
27, 28.	Cocoon and Chrysalis of Yellow-Bear Cater-	
	pillar	45
29.	Virginia Ermine Moth	46
30.	Crumple-Wing	49
31.	Salt-Marsh Caterpillar	49
32.	Arctia Acrea	52
33.	Acrea Moth (just out of its Cocoon)	54
34.	Arctia Acrea	55
35.	Chœrocampa Pampinatrix	60
36.	Chœrocampa Caterpillar with Ichneumon Chrysa-	
	lids	61
37.	Ichneumon Fly	62
38.	Ichneumon Chrysalids	62
39.	Chœrocampa Chrysalis	63
40, 41.	Cocoon and Chrysalis of Cecropia Moth	70
42.	Attacus Cecropia Moth	73
43.	Dryocampa Rubicunda	79
44, 45.	Caterpillar and Chrysalis of Dryocampa	
	Rubicunda	83
46.	Dryocampa Rubicunda Moth	85
47.	Saturnia Io (Female Moth)	86
48.	Saturnia Io Caterpillar, with the Three Grades	
	of Spines	88
49.	Chrysalis and Cocoon of Saturnia Io	90
50.	Female Io	91
51.	Male Io	92
52.	Macrosila Quinquemaculata Moth	97
53.	Larva of the Quinquemaculata Moth	98
54.	Chrysalis of the Quinquemaculata	99
55.	Ceratomia Quadricornis	101
56.	Caterpillar of Ceratomia Quadricornis	102

FIGURE		PAGE
57.	PHILAMPELUS ACHEMON MOTH	105
58.	CATERPILLAR OF PHILAMPELUS ACHEMON	106
59.	CATERPILLAR WITH HEAD WITHDRAWN	107
60, 61.	UPPER AND UNDER SIDE OF PHILAMPELUS CHRYSALIS,	108
62.	COCOON AND FRONT AND SIDE VIEWS OF THE ADONETA SPINULOIDES	110
63.	DIFFERENT POSITIONS OF THE BLACKBERRY LOOPER	120
64.	THE DRYOCAMPA IMPERIALIS	123
65.	CATERPILLAR OF DRYOCAMPA IMPERIALIS	125
66.	ATTACUS LUNA MOTH	129
67.	COCOON OF ATTACUS LUNA	134
68.	PAPILIO CRESPHONTES	137
69.	CHRYSALIS OF CRESPHONTES CATERPILLAR	141
70.	THE CRESPHONTES CATERPILLAR	142
71.	LARVA, PUPA, AND MALE MOTH OF THE PLUSIA BRASSICÆ,	146
72, 73.	HICKORY TUSSOCK MOTH AND CATERPILLAR	151
74.	COCOON OF HICKORY TUSSOCK MOTH	154
75, 76.	ORGYIA LEUCOSTIGMA MOTH (MALE AND FEMALE)	156
77.	CHRYSALIS AND WINGLESS MOTH OF ORGYIA LEUCOSTIGMA	157
78.	CATERPILLAR OF ORGYIA LEUCOSTIGMA	158
79.	CURRANT CATERPILLAR	163
80.	CURRANT SAW-FLY	165
81.	CURRANT LEAF EATEN IN CIRCULAR HOLES BY THE SAW-FLY	166
82.	THE BULRUSH CATERPILLAR	169
83.	LARVÆ OF NEW ZEALAND SWIFT MOTH	170
84.	ATTACUS CYNTHIA : EGGS, LARVA, COCOON, CHRYSALIS, AND FEMALE MOTH (AFTER RILEY)	178
85.	CATERPILLAR OF PAPILIO TURNUS	181
86.	PAPILIO TURNUS	183
87.	SCAPHA MOTH	193

FIGURE   PAGE

88. Hag Moth, Caterpillar, and Chrysalis   .   . 195

89. The Smartweed Caterpillar   202

90, 91. The Great Leopard Moth (Male and Female)

             and Caterpillar   207

92. The Eudamus Tityrus 212

93, 94. Caterpillar and Chrysalis of Eudamus Tityrus, 216

95. The White-Lined Morning Sphinx   .   .   . 217

96, 97. Caterpillars of White-Lined Morning Sphinx, 221, 222

98, 99. Apatela Americana—Moth and Caterpillar   . 223

100, 101. Young and Full-Grown Caterpillars of the

             Royal Walnut Moth .   .   . 231, 232

102. Chrysalis of Royal Walnut Moth .   .   . 233

103. The Royal Walnut Moth   235

"Oh look thou largely with lenient eyes
    On what so beside thee creeps and clings,
 For the possible glory that underlies
    The passing phase of the meanest things."
                                —*Mrs. Whitney.*

PREFACE TO "INSECT LIVES."

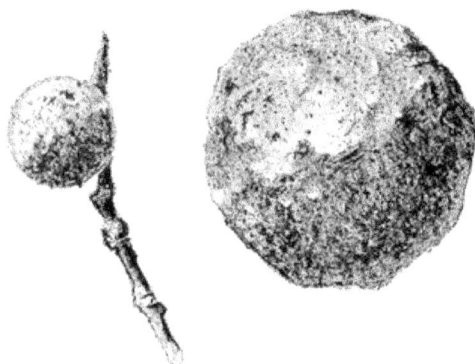

GALL-NUTS ("OAK APPLE").
HOME OF THE "CYNIPS CONFLUENS."

HOW shall we interest young people?
How shall we *most* interest them?
How shall we *best* interest them?

You give to your boy a glass ball. It is
clear and beautiful. He can amuse himself
with it. How? Not by studying it, but by
rolling or catching it. Tell him to put the

ball under a glass cover and watch it. Tell
him to wait and look again and see what he
will find. " Nothing," he says, " but a ball."
He is right. Man made it, and all the beauty
it will ever have it has now. Give him a
microscope. What does he see? A little
coarser texture, perhaps a flaw, a bubble of
confined air, but only the same glass ball.
Go with him to the forest. Pick from an oak
branch a plain brown ball. Is this *only* a ball?
Put it under a glass. Look again and you
will find it more than a ball. It is a home.
The doors will soon open and the family dis-
perse. Watch. There goes one in full dress
out on an early promenade. With what ease
and grace it walks up and down its prison of
glass. Another follows. There is a large
family for so small a house. Who built it?
Was it cast in a mould by a man? God made
it, and all the beauty it has is not seen at first.
Take the microscope. No roughness is revealed, no flaw, but exquisite beauty and finish
in every part of the house, in every part of
each perfect inmate. Suppose a boy could
buy a glass ball that would develop such wonderful secrets. What merchant could supply
the market? Aladdin's lamp would be at a
discount.

You give your girl a silk "beechnut-box."
Some of them will know what I mean: a
three-sided box, made of card-board and covered and lined with silk, such as only grand-
mothers can probably make now. She looks
at it. It seems solid. Press it and it opens.
One side has been left without being closed.
What can she do with it? It is better than a
ball. It will hold something. She can use it.
But the box itself, what will it come to? Tell
her to put the box under a glass and see what
it will get to be. She will laugh and tell you,
"only a box." All there is to it she sees at
once. Try the microscope. Only a little
coarser silk.

Here is a green "beechnut-box" I have
found on a walnut leaf. It is very small—no
larger than a beechnut and looking much like
a green one. Is it a box? Let us try the
microscope. It is embroidered on the sides
and back. There are small patterns in diamonds in brown and drab. While you look it
moves. Put it under a glass and watch. Is it
a home? Put a bit of walnut leaf by it. What
is that moving just under one of the pointed
ends? It is a head. The leaf begins to dis-
appear, the owner of the box, the *Limacodes
scapha*, is taking his breakfast.

Which will you prefer, the glass ball or the round, brown house, the silk box or the curious living thing that has surprised you and holds in reserve a still greater surprise ?

It is with the hope of getting this question answered in favor of living balls and boxes, of getting the key into the hand and getting the heart ready and anxious to unlock the many sources of beauty and interest which God has placed all about us in nature, that this little volume of " Insect Lives " has been written. That we may learn that while " it is the glory of God to conceal a thing," He is not only willing we should search out these hidden wonders, but will Himself be glad in our new-found delight in them.

EASTON, PA., Sept. 26, 1879.

LIMACODES SCAPHA.
(" BEECHNUT-BOX.")

PREFACE TO REVISED EDITION.

"AMONG the Moths and Butterflies again ! How long shall you study them, and enjoy them ? "

"As long as *you* study flowers and enjoy *them.*"

"But flowers are different. One *always* loves *them.* They are brighter and prettier in every way, and no care. *You* must be feeding and watching and waiting, always."

"And you must be planting and pruning, weeding and watching, waiting as long from seed to blossom as from egg to *imago.* Flowers are indeed bright and gay, but I do not admit they are brighter or more attractive than butterflies. As to feeding, you must at least give them *drink ;* and some, I learn, do not refuse food, nor object indeed to an occasional repast of beefsteak. Then the *life* of the moths and butterflies ! Your flowers, how-

ever fair and beautiful, are tied to earth. I
have heard, it is true, that some of them take
'a step a year,' and some are winged and
poised as if ready to fly, but *only* 'as if.' No,
they are servants to the butterflies. Holding
up dainty cups of ambrosia, leaving the lids of
their honey jars open, filling their chalices
with perfumed sweets at early evening, they
are the cupbearers to the floating fairies of the
garden, the meadow, and the wood."

So I *have* been out among them again, and
bring more stories of their triple life than I
gave in "Insect Lives ; or, Born in Prison,"
not forgetting those, but adding these to them,
wishing only there were more. Out among
them again and yet again, so long as the delicate
cups of the wild plum or locust entice the *Eudamus*, or the gay goblets of the tulip-tree shall
tempt the more brilliant *Turnus*, or the *White-
lined Morning Sphinx* hovers over the evening
primroses and four-o'clocks, and all the happy
children cry : "See, the humming-birds are
come ! the humming-birds are come !"

EASTON, PA., Sept. 1, 1890.

INTRODUCTORY.

" I KNOW you. I know what you *have* been. I know what you *will* be." This it is delightful to be able to say to the caterpillar crossing your path slowly, or to the butterfly winging its way in the air before you —to look upon the common brown brush-like caterpillar, with black at each end of him, and say : " Plod on a little longer, good fellow, and you shall be a tiger moth !" or upon the small yellow and white butterfly, and say : "You, a little while ago, were a green caterpillar, making holes for dear life through a cabbage leaf !" And this may easily be accomplished with the aid of your own eyes and a microscope, and also (as butterflies and caterpillars do not go flying and crawling about labelled) by the help of authors who have studied and classified them.

I. " But how shall we catch the butterflies ? With a net ? "

Not at all. That may do very well if you care for nothing but their present beauty ; but if you wish to know the butterfly, you had better take an earlier chapter in his life. Of course the first thing would be the egg, but, as these are not so easily found, you can begin with the caterpillar, and in due time you will came round to the egg, and so have the whole at command. The smaller the caterpillar when you get him the better, because he is very fond of changing his coat, and, liking a variety, is apt to put on quite a different one each time. Sometimes the second coat is much gayer than the first, even though *that* were a coat of many colors. Caterpillars usually change four times before going into a chrysalis state. Some butterfly caterpillars change five times (as the *Papilio philenor*), though the other Papilios of the Northern United States change but four, and some have but three changes ; so that one who has never noticed them carefully will be much surprised, in studying them, at the immense variety in shape and color, and also the great beauty which many of them display.

I have seen more elaborate work in design and color, in a surface less than an inch in length, and in width no more than a sixth of an inch, in a small, unnoticed caterpillar, than I have ever seen in as much surface on any flower. And the microscope reveals here often an amazing amount of work and beauty little suspected without its aid.

While some caterpillars are hairy, and look like little travelling clothes-brushes, others are knobbed, or spiny, like the porcupine, and others quite smooth. Some are handsomely dressed in scarlet and gold, with tufts of various colors grouped upon their bodies; and, strange as it may seem, some of the gayest and handsomest make the very dullest and homeliest moths. They have always *twelve rings*, called segments, besides a shelly head, and from ten to sixteen legs. They have a little conical tube or spinneret in the centre of the lower lip, from which they spin the silk for their cocoons, or draw the silken thread, which some use instead, to fasten themselves with when changing into chrysalids. The change from one coat to another is something curious, but not much in comparison to the change from the caterpillar to the butterfly through the chrysalis state. Here the *form* is entirely altered. The mouth and manner of eating and kind of food are totally different.

II. " But how can I touch the caterpillars when I wish to get them?"

Do not touch them at all. Take a little box, and when you see one, with a pencil or stick gently push him into it, and carry him home. Get some plain glass tumblers, the larger the better. You can begin with one or two, but you will soon want a dozen. Put your caterpillar upon a white paper, which you have first placed on an old book, or other firm substance, and cover him with the glass. If you have several kinds at once, it is well to label the glasses. Write "Grape," or "Apple," or "Poplar," upon a slip of paper, and paste it upon the tumbler which covers the caterpillar you found upon the grape, apple, or other leaf. This will avoid confusion, as they one by one go into chrysalids. You can study each one separately, and you will know, as they come out of the chrysalids (which you have seen them make), just which is the moth of the grape, apple, or whatever your label indicates should be there. This you would forget more easily than one would suppose.

You will thus know, also, at a moment's glance, how to feed them; as each caterpillar requires to be fed with whatever kind of leaf

you found him upon. If upon the grape, give
him grape leaves under the grape tumbler,
and so on. You will soon begin to respect
your caterpillar, and wonder at one thing
at least about him, and that is, his power of
selection. While there are a few, such as the
common salt-marsh caterpillar, that will eat
several things, as clover, plantain, and grass,
the most of them (at least so far as I have
tried them) will condescend to do nothing of
the kind. They know what they want, and
that is more than can be said of some people.
There is one kind of small caterpillar often
found on the grape-vine, and also on several
trees, which, although it prefers grape, will
eat other leaves ; but there are certain ones
peculiar to the grape, and you may try one of
these grape caterpillars with every other leaf
of the garden, and he will turn away with dis-
gust. Give him a grape leaf, and you are
paid for your trouble at once.

It sometimes happens that you will find a
caterpillar far from any tree or plant. Then
you can practice with him, and if you cannot
find out from a book what he is, and what he
should have, and fail to suit him with any
variety of leaf at your command, you must
either let him go, or see him die !

III. If you have very large caterpillars, such
as the elm, or royal walnut, or that of the
Polyphemus moth, it is easy to make a glass
box (bound with narrow ribbon, and fastened
at the corners), perhaps eight inches square
and six or eight high, or a box covered with
wire gauze. Such a box is better than the
round shades which you could buy, for you
can watch the insect much better through
them, and see it without distortion. It also
admits some air, which they require in order
to do well. It is needed for the large moths
also, which under a tumbler could not expand
their wings perfectly, much less make any use
of them. Here you can watch the caterpillar
dextrously fasten himself to the side of the
glass, and change his coat once, twice, three,
or four times, coming out each time fresh and
bright, and with a keen appetite after the stupid
supperless days each change costs him. You
can see him spin his cocoon with such a wonderful skill that you look with amazement at
the work ; or, if he changes into a smooth
chrysalis (as the Asterias butterfly), you can
see him fasten the loop around his breast,
which attaches him to the glass strongly
enough to keep him in one position (either
through a long or a short sleep), and at last

stand the tug of opening for the escape of
the butterfly. Besides this, if they are under
glass, they are safe, and you too are safe in
your knowledge of them. You know that
whatever living thing is found under your
glass when the chrysalis opens must have
come out of *that* chrysalis, whether legiti-
mately or not. The first Ichneumon fly I
ever examined would have been brushed un-
ceremoniously out of the window for a wasp,
had he stolen out from an unguarded chrysalis. But, as he was born in prison, there he
was. He came out of *that* chrysalis, and wasp
or what not, he must be studied, and lo ! the
curious parasite was brought to light. Revela-
tions of this kind will sometimes be made,
which one would be slow to believe possible,
but for there being, in this way, no possible
room for doubt. I have had two caterpillars, for example, which were just alike, spin
each a cocoon exactly alike, each being under a glass of its own and labelled. After
a time, on cutting open the cocoons carefully,
so as not to injure the chrysalis (which may
be easily done), one cocoon was found to con-
tain a perfect chrysalis. The other contained
the dead caterpillar and four rather small oval
chrysalids. Finally, the *one* perfect chrysalis

opened for the escape of a moth (*Apatela
americana*), and the other four small chrysalids opened, and lo! six large flies, much
resembling the house fly, only more spiny or
hairy. There must have been two flies in
two of the cocoons, as there were certainly
two extra ones under the glass!

IV. The immense variety of caterpillars,
and the great difference in their habits, and in
their new and finished life as moth or butterfly, furnish constant surprise and pleasure in
their study. From egg to *imago* (which means
the perfect insect or butterfly) they are a study
which cannot fail to excite wonder, and lead
us, from admiration of their beauty and skill,
to adoration of Him whose work is perfect
though invisible, and whose ways, studied
never so closely, are still "past finding out."

To render our researches most effectual and
satisfactory, we should not begin with statis-
tics—studying how many thousands of moths
and butterflies there are supposed to be, or
how many species of insects have been classified and named. Take "one to begin," as
children say, and study it thoroughly. From
books such as those of Edwards, Harris, Packard, or Tenney, find the name of your caterpillar, and know, before he changes, what sort

of butterfly you are to have; unless you are
fortunate enough to find one not described,
and then you can have the honor of naming
him yourself. In this way the more scientific
knowledge to be obtained from books you
will soon find it impossible to do without.
You will find that while it is pleasant to be
sent from books to nature, it is more pleasant
to find out secrets from nature, and let her
send you to the books to verify them.

V. But there are a few things you should
know from books before you begin, and one is,
that the whole class of butterflies and moths
is called LEPIDOPTERA ; and that this class
contains only Butterflies, Moths, and Hawk-
Moths. Flies, beetles, and other insects come
under different classes.

The *Butterflies* have delicate thread-like
antennæ, and these are always knobbed or
thickened at the end. They always fly by
day, and their caterpillars have sixteen legs—
six small tapering, jointed ones (which are the
true feet) from the first three rings back of
the head, and a pair of larger and more fleshy
legs to each of the other segments except the
fourth, fifth, tenth, and eleventh.

The *Hawk-Moths* have long narrow wings,
and some of them look very much like little

humming-birds. Their antennæ are tapering
(usually broader in the middle), and never
knobbed. They fly rarely during the day,
but mostly in the morning and evening
twilight.

The *Moths* have not narrow wings. Their
antennæ are not knobbed but usually taper
from base to tip, and are not broader in the
middle like those of hawk-moths. Some of
them are spined and some plumed. They fly
at night chiefly. So you can always tell a
butterfly from a moth by the antennæ, and a
hawk-moth from a moth by its wings.

The eggs are very different in size, shape,
and color. Some are clear and round like little
crystal beads, and formed on a leaf in a close
circle. Sometimes they are in exact rows and
of an amber color. Again, like those of the
Polyphemus moth, they are chocolate-colored,
circular, flat, and quite large. The eggs of
this moth are shaped like biscuit, and have
two white rings around the edge. Some eggs
now before me, found to-day in a walk to the
woods, and unknown to me, are white, as if
made of milk glass. They are on a large
forest leaf, and there are just ninety-one of
them, and yet I could cover the whole with a
thimble. They look like plain " chalk beads,"

and may be easily counted with the naked eye, but look at them through a microscope, and their exquisite beauty appears. They are all precisely alike, having sixteen or eighteen symmetrical grooves diverging from a small circle in the centre like this : <small>Fig. 1.</small>

And what is more wonderful than the finish of the egg, is, that the different kinds of eggs are always placed upon that kind of leaf, which, when the caterpillar is hatched, he will at once prefer to eat, except, of course, those you may have in your box, or under your tumbler, and then you will know what to feed them. But, as I said, the best way is to begin with the caterpillars, as you will seldom find the eggs in any other way, or have success in raising such, if you should.

VI. *How to kill a moth or butterfly.* Butterflies and moths having so much vitality, it has been a puzzle how to kill them without injuring the delicate texture of their wings and without pain. A sure and easy way is the following :

Take a glass jar with large mouth and close lid (a candy jar, six inches high and four inches diameter, with glass cover shutting over a rubber band is good), into which put four or five lumps of cyanide of potassium about the size of a hickory-nut. Dissolve enough plaster of Paris in water to cover the cyanide evenly over, forming a hard smooth surface. Put the moth into the jar, close the lid and let it remain five or six hours, after which it can be taken out and mounted.

Have a board (smoothly planed) with a groove the size, in length and width, of the body of the moth. Place it upon the board with the body in the groove; spread the wings evenly, and confine them by strips of paper placed across so as to hold the border of each wing. Take off the papers the next day, and with a pin through the thorax, fasten it to the cork gummed upon the box in which you place it.

" The velvet nap which on his wings doth lie,
   The silken down with which his back is dight,
   His broad outstretchèd horns, his hairy thighs,
   His glistening colors and his glorious eyes."
                                    —SPENSER.

AMONG THE MOTHS AND
BUTTERFLIES.

I.

BORN IN PRISON.

I AM only a day old! I wonder if every butterfly comes into the world to find such queer things about him? I was born in prison. I can see right through my walls; but I can't find any door. Right below me (for I have climbed up the wall) lies a queer-looking, empty box. It is clear, and a pale green. It is all in one piece, only a little slit in the top. I wonder what came out of it. Close by it there is another green box, long and narrow, but not empty, and no slit in the top. I wonder what is in it. Near it is a smooth, green caterpillar, crawling on the edge of a bit of cabbage leaf. I'm afraid that bright light has hurt my eyes. It was just outside of my prison wall, and bright as the sun. The first thing I remember, even before

my wings had opened wide, or I was half
through stretching my feet to see if I could
use them in climbing, there was a great eye
looking at me.   Something round was before
it, with a handle.   I suppose it was a quizzing-
glass to see what I was about.   I heard some-
body say, " Oh ! oh !" twice, just as if they
wondered I was here.   Then they held the
great bright light close to the wall, till my
eyes were dazzled.   I don't like this prison.

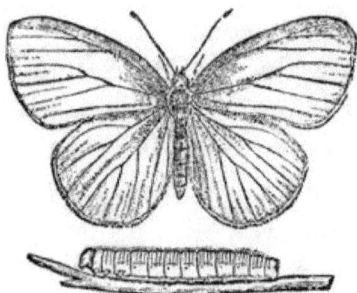

FIG. 2.   FIG 3.
CABBAGE BUTTERFLY AND CATERPILLAR.   CHRYSALIS.

It is n't worth while to fly about.   It seems as
if I ought to have more room.   There must be
something inside that green box.   It moves !
I saw it half tip over then, all of itself.   I
believe that caterpillar is afraid of it.   He
creeps off slowly toward the wall.   How
smooth and green he is !   How his rings
move when he crawls !   Now he has gone up

the wall. He has stopped near the roof.
How he throws his head from side to side!
He is growing broader! He looks just as if
he was turning into one of these green boxes!
How that box shakes! There, I see it begin
to open! There is a slit coming in the back!
Something peeps out! A butterfly's head, I
declare! Here it comes—two long feelers,
two short ones! Four wings, two round spots
on each of the upper pair, and none on the
other two. Dressed just like me. I wonder
why it hid away in that box?

First Butterfly.—" What made you hide in
that green box?"

Second Butterfly.—" What box? I have n't
hid anywhere. I don't know what box you
mean."

First Butterfly.—" That one. You just
crawled out of it. I saw you."

Second Butterfly.—" That 's the first I
knew of it. There are *two* boxes, just alike.
Both empty. May be you were hid in the
other!"

First Butterfly.—" Ho! There goes up our
prison wall! That 's the big hand that held
the bright light. How good the air feels!
Now for a chance to try our wings! Away
we go!"

II.

THERE is a very pretty caterpillar which lives upon the common milk-weed, or *Asclepias*, which grows by the roadside, with pinkish clusters of flowers in summer, and curious bird-shaped pods in the fall. This caterpillar (whose name is *Danais archippus* —we might call him Archie, for short) is very pretty, and the butterfly is handsome ; but the crowning beauty of all is the chrysalis. It looks like a little green house, put together with gold nails. It is somewhat of the shape and size of a long, delicate pea-green acorn, and has a row of dots half-way around what would be the saucer of the acorn, with others about the size of a pin's head on different parts of the chrysalis, and you will say they are not like gold, but are real gold itself.

The caterpillar, when full-grown, is about two inches long. It is cylindrical, and hand-

4

somely marked when mature, with narrow alternating bands of black, white, and lemon-yellow
(Fig 4). These bands are not entirely even,

FIG. 4. DANAIS CATERPILLAR.

and occasionally run into each other. On the
top of the second ring, or segment, are two
slender, black, thread-like horns, and on a hind
ring two more, not quite so long as those near
the head. You can find it almost any day in July
or August, if you look closely,
on the underside of the broad
ovate-elliptical leaves of the
milk-weed. When this caterpillar first leaves its conical, re-
ticulated egg (Fig. 5, which
is always found on the under
side of the leaf, a miniature
hanging basket, first yellow
and then gray, as it devel-

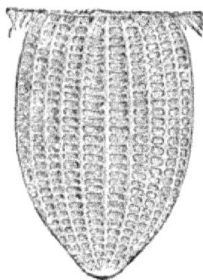

FIG. 5.
AN EGG, MAGNIFIED.

ops), it is perfectly cylindrical, and of nearly
the same size throughout, and only twelve

one-hundredths of an inch in length.   In this, its *first coat*, it is a pale, greenish white, and the horns (front and back) are mere conical points, and it is covered with little black hairs or bristles, from minute warts on the back and sides.   The breathing holes, or *stigmata*, show on each side, marked by a plain, narrow band.   In the next coat, which it puts on in a few days, the black stripes appear, and also faint lines of white and yellow, and the horns are longer.   The third and last coat (before the final change to the chrysalis) is much the same, except that all the colors are brighter.   The horns are shed with the skin, new ones having been formed beneath to take their place.   These have been so carefully folded away that at first they scarcely appear ; but they are soon developed, or uncurled, and

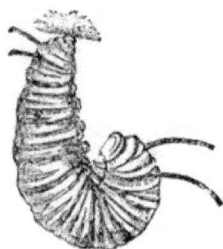

FIG. 6.

unbend so suddenly as almost to surprise one.

When the caterpillar is ready to make its change into the chrysalis, it spins a little tuft or button of silk to the under side of the leaf (or the box-cover, if in prison), into which it fastens its hind legs, by their little hooks, then lets go the hold of its other legs,

and hangs, head downward, with the body
curved, as in Fig. 6.

In this position it remains about twenty-
four hours when the marvellous
change is wrought—the coat
thrown off and the chrysalis
(Fig. 7) developed.

It was the accidental finding of
this chrysalis, attached to a spray
of wild carrot, that led me to
study this particular species.   It
was a secret to me—this beautiful
green-and-gold house.   It held

FIG. 7.
DANAIS CHRYSALIS.

something. What, I must know! Cutting the
stem of the carrot, I brought the treasure carefully into the house, covered it with a tumbler,
and for a week it remained just the same. Then
the green began to turn to a light purple, and

FIG. 8.   FRONT VIEW.   FIG. 9.   BACK VIEW.

lines began to show through the clear case.
The front showed lines like a curtain, parted
and folded back each way, like drapery to the

bottom, as shown in Fig. 8. The back was curiously marked off, and looked like Fig. 9. The whole gradually took on a very dark purple hue, and I hoped to see it open and give up its treasure. But though I watched very carefully, it stole a march on me, and one morning I found its secret disclosed and fluttering below the empty chrysalis, now but a clear, rent tissue, with here and there a pale gold dot.

The butterfly is handsome and quite large (more than three inches across when the wings are spread), but not quite so beautiful as you would infer from his elegant house. He is of a rich tawny orange, bordered with velvety black on the upper side, and a lighter nankeen yellow below ; and has a large velvety black head, spotted with white.

As I did not know how large he would be, nor when he would come out—for he did not invite me, as I said, to his "opening,"—I had not given him a glass roomy enough for his wings to expand entirely at the first, as they must, or remain imperfect. So afterward, although he had the liberty of the whole room, he walked about with one wing folded back over his shoulder, like a lady's opera-cloak. But I kept him, and, learning that he came from the

milk-weed caterpillar, I went in quest of one.
I was fortunate enough to find five in one
search—three on one milk-weed, and two on
another. I put them in a glass fernery, about
one foot long and ten inches high, and fed
them with fresh milk-weed leaves daily. Soon
they mounted, one after another, to the top,
and began to work on the under side of the
glass cover. My curiosity was on the alert
to see how each would build his green house.
I had seen cocoons of various kinds spun;
but the glass-smooth chrysalis could not be
spun. ‚Oh, no! It was altogether too nice
work to be done in sight. There was no
sound of hammer or sight of tools. It was all
polished and painted and ready—and lo! the
inner layers of the caterpillar's skin had been
the workshop, and the outer skin was taken
down and discarded, like worthless scaffolding,
when the green-and-gold house was ready.
Pretty soon there were five of these houses
hanging from the glass roof, side by side; and
now there are five empty homes still clinging
by the little shiny black twist that fastens
them firmly to the glass, and five handsome
great butterflies, like the one shown in Fig.
10. Only one of all these did I see break the
shell and come out, and that only by the most

diligent watching. The butterfly was packed, head downward, at the bottom of the chrysalis —wonderfully packed, as all will admit who see him emerge, to shake himself out into something five or six times as wide, a beautiful uncramped butterfly.

FIG. 10.   DANAIS ARCHIPPUS.

After seeing them brighten a bouquet, and watching them eat with their long spiral tongues from a little bed of moss sprinkled with sweetened water, I let them take a nap under a tumbler with a little pillow of chloroformed cotton, and, unmarred even by a pin, they were ready to be laid away in a glass-covered box in their long, dreamless sleep.

It has been said by some entomologists that each plant is visited by about five different insects. This year (1877) I have searched in vain on the milk-weed for the large, handsome caterpillar of the *Danais archippus.* That there must have been a few the occasional presence of the Danais butterfly has proved. Two were seen in Massachusetts, flitting gayly past me as if in mockery of a long and futile search I had just made for the caterpillar among a whole tract of milk-weed ; one in Brooklyn, and one or two in Pennsylvania, but they were exceedingly rare. The eggs were probably destroyed by spiders and other insects, but why to so much greater extent than the previous year is not so readily explained.

The only caterpillar (and that very abundant) which seems to have lived upon milk-weed this year, and found upon the same spot where the Danais caterpillars were so readily obtained last year—sometimes half a dozen upon one plant,—is a small one in comparison to that of the Danais, of a soft, woolly appearance, orange-red in color, and about an inch in length, with hairs thickly set in starry clusters about each fleshy ring. Three of these abundant orange-red caterpillars have

gone into shiny-brown chrysalids and come
out, after a three weeks' sleep, into lavender-
colored moths, perhaps an inch and a half
across the expanded wings, the wings edged
with a narrow orange border. They were
"travelled" caterpillars, going in a box as
chrysalids from Pennsylvania to Massachusetts, coming out there, and travelling back as
quietly as if long journeys were a matter of
course. A second set of caterpillars of the
same kind appeared in August, some of which
are now (September) in their chrysalid homes.
They made from their woolly, downy hairs
(more soft than those of any other caterpillar
I have seen) a soft cocoon like loose felt, and
these four have gone up in pairs, two chrysalids in each thin cocoon. This little lavender
moth is neat and quite pretty, but not to be
compared for beauty to the *Danais archippus.*

It has always been with a feeling akin to
sadness that I have seen the walls of the
beautiful home of the Danais butterfly break,
and its beauty vanish, even for the release of
the scarcely less lovely winged creature that
sails off, regardless of its shattered home. It
is not so strange after all that it should be
able to leave it without regret, when one con-
siders that no Danais butterfly has ever seen

the handsome house it lived in ! For before it can escape the walls grow very thin, the gold nails vanish, and when the rich brown and orange-yellow butterfly steps out so airily, there is nothing left but a clear bit of broken glass-like material to hint of the once exquisite green and gold home.

But *now* the butterflies can see what sort of a home they had, if not their own, those of their neighbors, precisely like them. Here are green houses, as perfect after more than their usual fortnight has gone by as when first made. The gold nails still bright, and the walls intact. The butterfly has been requested to *stay at home;* and if he had any objections, they vanished so soon as his house was placed in that safest of all places, the cyanide jar ! [1]

Five of these houses (a very handsome block) I have now before me (September, 1889), in a row, to remain permanently ; with the satisfaction of knowing that the imprisoned occupant can never realize what it has sacrificed for my pleasure, in thus *staying at home.*

Making a collection of the *eggs* of butterflies

[1] The arresting of the transformation of the Danais, by placing the chrysalis for some hours in the cyanide jar used for killing the perfect insect, was a new thought to me, which experiment proved a success ; and which may open the way for the preservation of all chrysalids.

and moths (or of any insects, in fact) is only
second in interest to the collecting of the
perfect insects themselves. And this is far
more easy than one would suppose. Looking
on the under side of a forest leaf, or of a plant
or vine near your door, will often reveal clusters of eggs, that one not "on the search"
would never dream of being there. The last
summer I secured more than two dozen eggs
of the *Danais archippus* by searching the
leaves of the milk-weed; never finding but
one on a leaf, and that one always on the
under side of the leaf, and so small as to
escape notice but by a careful and practised
eye. The egg is of a light color, and about
as large as a "period" in the book you are
reading. On the 6th of August last (1889),
I watched one of these tiny eggs open, and I
shall never forget the pleasure I experienced
as I saw the little prisoner make a minute hole
in the egg and put out a jetty black head,
turning it this way and that, before he left his
prison, as much as to say: "I wonder what
sort of a world it is that I am about to step
into!" He was not long in deciding the
question "Is life worth living?" and bravely
stepped forth to try it. I noted this as the
greatest amount of intelligence in the smallest

compass that it had ever been my good-fortune to witness! He tried "life" and found it, with a plentiful supply of milk-weed, well worth living, went through all his changes till he entered his royal castle of gold, and came forth to a higher life, which, as long as it lasted, was only one of unmixed enjoyment.

On another leaf, from the maple tree, I espied seventeen glassy, bead-like eggs, and from them came seventeen of the beautiful Rosy Dryocampas, now waiting in their notched chrysalids their time of winged freedom.

Upon a maple leaf on the tree, and upon a pretty high bough, I espied, last fall (September, 1889), in walking by, what I at once divined to be the egg of a Polyphemus moth. Securing it, in spite of the smile of the friend with me, who thought it impossible to see an insect's egg of any sort at that distance, much less to determine its character, I am rewarded whenever I look at the fine large Polyphemus cocoon, now almost ready for its spring opening. Other clusters of eggs, larger in numbers than those named of the Dryocampa and Danais were found, some of most exquisite finish and beauty. The idea given above about "*jarring* the chrysalids," was transferred to the *eggs*, and by placing such a

portion of any cluster found as I did not
wish to have *try life*, in the cyanide jar, I
found them ready to place in a box in my
insect-egg collection.

FIG. 11.

III.

A BUTTERFLY in March! Velvety black, with wings bordered with a double row of yellow spots, and the hinder wings tailed, having also the added ornament of seven blue spots (a nebula of dotted blue points, with a frosted silvery sheen marking each spot). He is the *Papilio asterias* (Fig. 13). You have seen him in May, June, or July, hovering over a bed of phlox or other sweet flowers; but unless you caught him "in the bud," or, of course, when a caterpillar, you would not have him in the middle of March.

FIG. 12.   CATERPILLAR OF PAPILIO ASTERIAS.

The sole occupant of a glass fernery, sipping from sugar-sprinkled moss with his long, un-

2   17

coiled tongue, he seems quite at home, and sees nothing of the snow now whitening every branch and tiny shrub—knows nothing of the "April-fool," which, as Susan Coolidge says, spring throws to the flowers outside—the daring crocus and daffodil. With his moss, and some fresh snowdrops in a vase, standing in his glass house for dessert—an extra drop of sweetened water in their pure cups—he is monarch of his little world.

As a caterpillar, he was handsome. At first a tiny black caterpillar, with a white stripe running through the centre of the body and across the tail, and covered with some small black dots or points. The next coat has but one white stripe across the middle, on the sixth and seventh rings, with orange spots beneath the black points, two white spots on his first ring, and a row of white spots on each side. Then at last he has a rich coat, striped with black and dark green, and ornamented with deep yellow spots (Fig. 12). But his chrysalis is quite plain, with nothing of the exquisite beauty of the green- and- gold house of the Danais. But when he leaves his shell, coming out by the narrowest possible front door, so that you must look sharp to see the thread-like opening, then he is much handsomer than the

Danais butterfly. So, many people, living in plain tabernacles, and sometimes regarded homely by others, have something within, waiting to give great surprise, when they shall have escaped, through a narrow door, into a world of wonderful light and beauty !

The *Papilio asterias* is very fond, in his caterpillar form, of the wild carrot, or garden

FIG. 13. PAPILIO ASTERIAS.

carrot, parsley, or celery, and any of the warm, aromatic plants, as anise, caraway, and dill.

This March butterfly, as a caterpillar, was eating his delicate carrot leaves and seeds last September at the same time with the Danais caterpillar, and as we brought them fresh leaves, day after day, and watched them go into their queer little houses at the same time, we did not know then but they would have

their " opening " also, together. But while the
Danais was ready to come out in a fortnight,
or three weeks, the Asterias slept on until
March—six months under his glass roof, without moving a hair's breadth, until he was out
trying his new wings yesterday morning.
Some other kinds of chrysalids have kept him
company all this time, except that they have
moved a little, and sometimes a good deal
(when touched with a pencil, or slightly blown
upon), showing the life within ; but not a
particle—watch him never so closely—moves

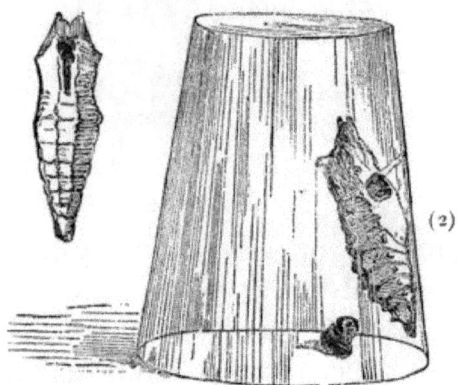

FIG. 14.

the Asterias. There were six chrysalids of
this one kind under separate glasses ; all of
which were taken as caterpillars, and each of
which I had watched go into his separate

house. It is not a cocoon, woven as some are of
their own hairs, or spun from some hidden substance through a spinneret; but like the Danais'
it is formed under the caterpillar skin, and
when he is suspended as a caterpillar, with a
silken thread holding him about the body, as
shown in the picture (Fig. 15),
he drops off the entire skin,
and it remains, as seen, beside
his chrysalis, which is pale and
nondescript in color, knobbed
with many little round pro-
tuberances, giving it a curious
rather than pretty appearance.
When one was out, the next

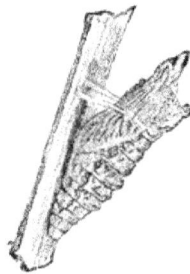

FIG. 15.

thing was to look at the others, when lo! a most
surprising revelation! Another chrysalis was
empty, but the *front door* was very different!
Instead of a crack, a thread wide and half an
inch long, in the upper part of the back (Fig.
14, 1)—(the narrow black line in the chrysalis
shows the butterfly's door), there was in the
side (marked O in the picture of the chrysalis,
and only belonging there to show this second
front door) a perfectly round hole (Fig. 14, 2),
the size of a pea ; and trying his new wings (four
narrow, glossy, blue-black ones), was something
more unlike the butterfly than was the circular

door he came out of unlike the narrow door of
the Asterias.   Looking something like a saw-
fly, and more like a wasp, it was a large ichneumon fly.   The parent ichneumon, having stung
the caterpillar and deposited the egg, the
ichneumon was safe in his provided chrysalis
home, when he woke up to a sense of his priv-
ileges, and not only appropriated the house of
the Asterias, but literally lived on the occu-

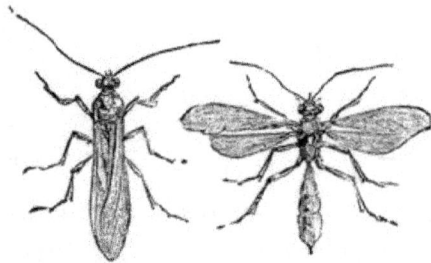

FIG. 16.   ICHNEUMON FLY.

pant, eating him up and then making his own
way into the world, leaving the chrysalis
entirely empty, and quite whole, with the
exception of the round door.   His head and
slender body, antennæ, and six feet, are all an
ochre yellow.   The eyes are large, jetty black,
and oval-shaped, and back of them, on the top
of the head, are three round, black beads, in a
triangular position.   His body is joined to his
head and shoulders by a pedicel, so long and

slender that he is able to work from it like a
pivot, in all directions, giving as fine specimens
of gymnastic operations as one often sees.

His veined, clear wings are exquisitely
glossy, and he polishes their steel blue till
it burns like a mirror. He has the vanity of a
Beau Brummel, judging by the great pains he
hourly takes with his entire toilet. Grasping
both his long trembling antennæ at once, and
smoothing them out again, as a philosopher
would stroke his beard, nothing is left on one
of their thirty-five segments large enough for
a microscope to reveal. Then his wings and
six legs go through the same operation, and
he is ready for a fresh supply of sugared
sweets. But alas, his mouth ! If he had claim
to beauty in every other particular, one good
look at this remarkable feature in a mirror
would secure his humility for ever. An hour's
close study with the microscope reveals no
trace of beauty about it ! The most curious
transformations do no good in redeeming its
unmistakable homeliness. There are three
projections from it—impossible to describe—
two seem like short, curved legs, with which
it clasps its throat, and the centre is a curved
affair something like the letter V. It is
very much like the mouth of a wasp, but in

such constant motion that one cannot guess at its exact shape or manner of manipulation.

It is well that it is so small that it does not detract from his looks except with the use of a microscope—and so long as he does not know it himself we will allow his vanity to be pardonable.

One such parasite will, however, satisfy us, and we hope only the *narrow* front door will open for the rest of the Asterias chrysalids.

IV.

THE EARLY BUTTERFLY.

WALKING up a rocky lane one warm day in the latter part of winter, my attention was called to a large, sombre-looking butterfly, lying flat upon a rock. Any sort

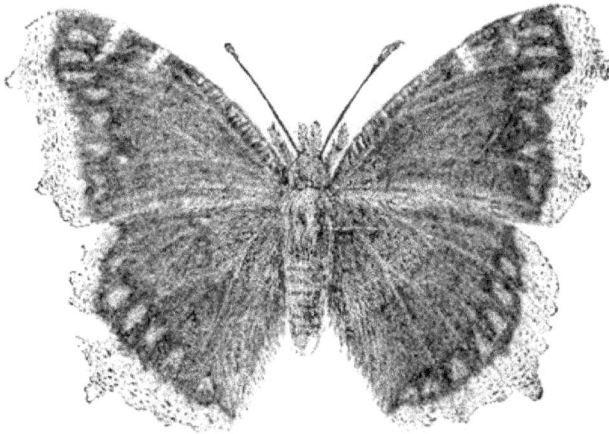

FIG. 17. THE EARLY BUTTERFLY. VANESSA ANTIOPA.

of butterfly, so out of season, was worthy of notice, and as this one was very quiet, as if half asleep, I easily took him up and carried him home with me. He was handsomer upon inspection than at first sight I had imagined. The wings, though grave in color, were really a rich purple brown, with a broad margin of light yellow or buff, and six or seven spots of a lavender color inside of the border on all the wings. He had a queer, pinched-looking head, with sharp features, and furry front feet. I did not know his name, and as he was very restless, and beat constantly against his prison wall, I gave him his liberty. Some months after, on June 5th of the same year, I found on a shrub, in the same rocky lane, a very formidable-looking spine-covered caterpillar (Fig. 18).

FIG. 18.   VANESSA ANTIOPA CATERPILLAR.

He was black, but dotted with minute irregular white spots, like tiny snow-flakes. There was a broad black line running down the back, interrupted by eight spots of brick-red. Each side, also, was dotted with white spots. There were seven rows of large spines, besides a row of very small but similar ones low down, just over the feet. Each of the

two centre spines on the ten rings were
branched, as also the two on the last ring.
As these spines were stiff and sharp, and did
not lie particularly close to his body, he was
treated in a very cautious manner until safe
in his glass prison, although I have been told
that these caterpillars, and in fact nearly all
caterpillars, however formidable they may
look, are in fact harmless. The fiercest one
I have ever seen, that of the regal walnut
moth (*Ceratocampa regalis*), very large, and
with horny spines stretched over the head,
which when disturbed he shakes in a threaten-
ing manner, is said to be perfectly harmless.
One would certainly prefer to test this harm-
lessness when he had thrown off his horns,
and, after a smooth, chrysalis life, come out
into the beautiful walnut moth.

The caterpillar I had imprisoned did not
at first like his confinement at all, and showed
a most worthy persistency in attempts to solve
the possibilities of escape, walking with entire
contempt over the fresh leaves of the willow
from which he was taken (and any species of
which he will eat), going up and down and
across to every corner and joint of the box,
until, at last, apparently satisfied that he was
secure in his new abode, he wisely accepted

the situation and began such a marvellous course of eating as showed that he had determined, if he must be a prisoner, not to commit suicide by starvation. Leaf after leaf disappeared and new ones were supplied, until, at length, he suddenly stopped eating, and began to weave a little thread and fasten himself securely at right angles with the side of the box, much in the same way as the Danais caterpillar. His head is round, large, and flat on the top, resembling the old-fashioned velvet " jockey cap." There is no red spot on the first two rings from the head, but on all the rest ; each spot, on close examination, being made of three spots close together in the form of a triangle, in this manner .˙. Nothing could be much meeker, or in greater contrast to his first eager restlessness and snappishness, than his appearance after he has fastened himself by his hind feet firmly to the glass, with his head downward and bowed forward touching the glass, only a slight movement of the head now and then showing that he is alive. His three pairs of true feet he draws close together like a wedge, in short spasmodic movements, and then slowly opens them again. At last, after a day or more of this suspension, he throws off the caterpillar skin and shakes himself into a brownish chrysalis, which operation takes but a few seconds after it has begun. But the chrysalis, which at first is soft and misshapen, has to assume its characteristic form, which it does by contracting and expanding and throwing out a protuberance, until, in about an hour, it has its shape, and its surface becomes hardened and the chrysalis complete.

FIG. 19.
CHRYSALIS
OF VANESSA
ANTIOPA.

This was on the 6th of June, and on the 18th day of the same month the chrysalis opened, and lo ! there was my early winter butterfly, the *Vanessa antiopa.* This one was much fresher and prettier than the one found in February, and this I could well account for when I learned that this butterfly lives often all winter, hiding in some sheltered spot, stupid and almost dormant, but ready for the first sunny day, sometimes enticed from its hiding-place before the snow is quite gone, its wings somewhat worn and faded by its winter's experience. Since then I know it as the earliest butterfly, and am not surprised to see it early in February heralding the spring far in advance of any other.

V

THROUGH A GLASS CLEARLY.

WE do not like to see a beautiful thing at a disadvantage. When a large co-coon (Fig. 20, yellowish-brown and leaf en-wrapped), cut from a spray of wild raspberry, in September, had been watched for over six months, and showed no signs of life within, it was half given up as a useless affair. Inquiring

FIG. 20. POLYPHEMUS COCOON.

scissors, one day in March, stole an entrance into the cocoon by carefully snipping one end, and cutting spirally round an opening which revealed, unharmed, the living chrysalis within

(Fig. 21). It seemed certain—secret as it then was—that from out this brown-ringed casket some beautiful thing was preparing to emerge.

FIG. 21. BACK VIEW.

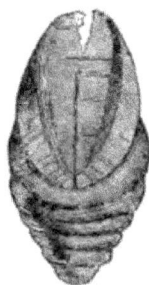

FIG. 22. FRONT VIEW.

While watching it closely, a month later, one of the vest-like folds on the breast (Fig. 22) slowly began to part, revealing, first, a curious bridge of fringe across the opening. What could this be? The side of the clear-glass box, even, was too much obstruction for the impatient watcher. "I cannot look at this through a glass darkly," I said, as the lid was removed; and slowly out came this amber fringe, a broad, beautiful antenna, yellow stemmed from base to tip, with ochre-yellow fibres radia-

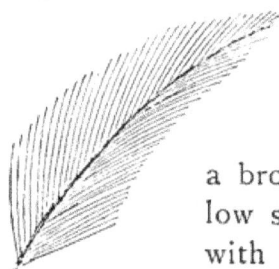

FIG. 23. ANTENNA. ting from it in a perfect plume. The other soon followed. So large, so full, so beautiful antennæ I had not seen before.

Now for the microscope. Ah, the difference between an obstructing and a revealing glass ! Between seeing through a glass darkly and through a glass clearly ! A richly-colored centre stem, of thirty-one joints, and two fila- ments to each joint, of exquisite finish and symmetry. Then a little wider parting of the vest (no breaking of the chrysalis), now and then a shiver and a spasmodic movement of the whole chrysalis, with a little further exit— another shiver, another waiting, and in an hour and a half out came a beautiful (but still limp and contracted) Polyphemus moth (Fig. 24).

A pot of hepatica stood ready in the box for him to cling to while expanding his wings, but the slight, fresh stems proving too frail for his weight, the danger of a fall was pre- vented by putting a stick into the earth beside the hepatica, to which he immediately clung, and gently unfolded his soft-hued ochre wings, bordered with gray, showing two large and elegant eye-spots on the hinder ones, of a deep blue-black, with a transparent oval in them, clear as a bit of inserted mica. In the upper wings were two smaller transparent ovals ; a collar, edged with lake color, and two spots of lake-red, edged with black on the edge of the upper wings, completed his beauty. The body, a soft brown ochre, was furry and

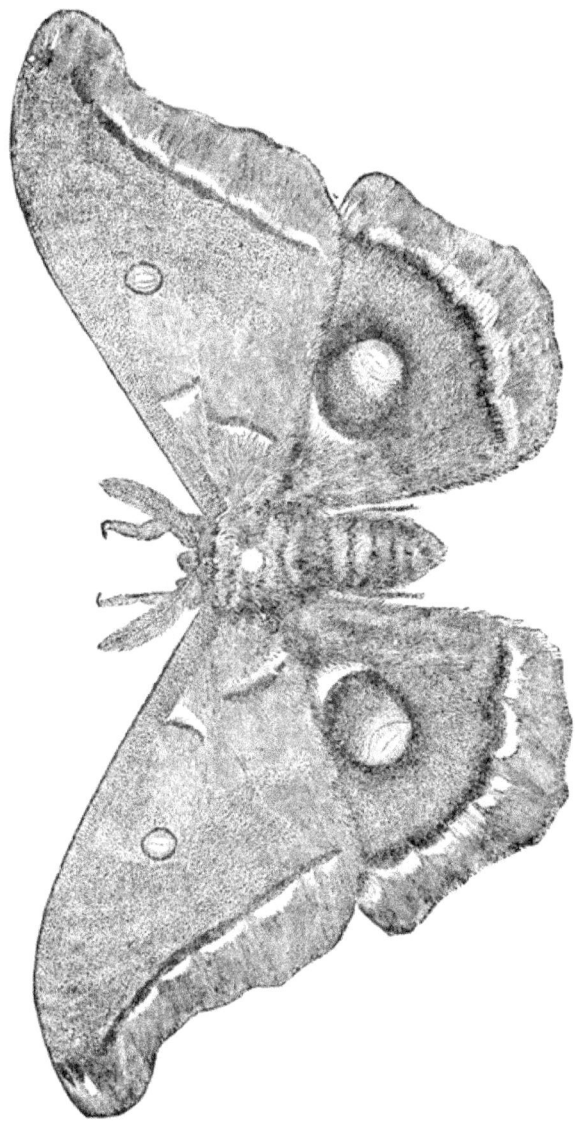

FIG. 24. THE POLYPHEMUS MOTH.

33

feathery as an owl. Large eyes, six short furry dark-brown legs, a softness of blending in color, and a gentleness and grace of motion crowned the whole. Lifting his large wings, his flight was slow and graceful ; no hurried fluttering and wild beating against the glass when a prisoner ; no dashing about the room when at liberty.

If ever a name was a misnomer, it is surely so in his case. Polyphemus, a one-eyed furious giant, a murderer and greedy cannibal, for *him* to give a name to this two-eyed, gentle-natured, and apparently tongueless moth (whom no sweets could tempt), simply because it is large ! As well might he be called the Tower of Babel, Behemoth, Leviathan, or any other great thing of earth or sea. He is, however, not likely himself to apply to the legislature for redress for this grievance.

The inside of this cocoon is finished with the hardness and smoothness of the inside of an almond shell which it closely resembles, except being much larger.

The larva of this moth is described as of a bluish-green color, with a yellowish-brown head, living upon the oak, elm, and lime trees ; the cast-off skin was enclosed in this cocoon. The disposition of the eye-spotted ogre was

well tested in the artist's saloon. No philosopher ever showed more patience and dignity
under repeated trials at the hands of a pho-
tographer than he displayed in the hands of his
persecutors, with no knowledge of the cause to
stimulate his vanity and inspire his courage.

I said the mystery wrapped up in the brown
cocoon was "a secret." In studying Natural

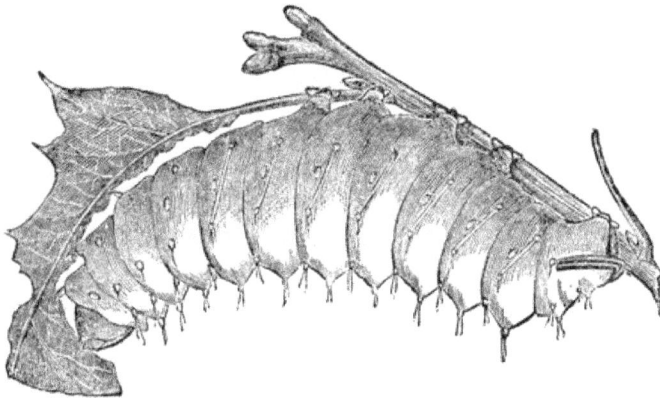

FIG. 25. POLYPHEMUS CATERPILLAR.

History we often learn the first part of a lesson
last; sometimes the middle part first; sometimes it is years after we get part first before
we can find part second, even of a short, small
lesson. The pages of nature's book are count-
less, but they are not all numbered, and sometimes we have to stop and wait in a most
interesting place. It is all the pleasanter when

we complete the round.   After the Polyphemus moth had been mounted for months, a
beautiful caterpillar was given to me (Fig. 25).
He was very large ; of a handsome pea-green
color, with little points of golden yellow, which,
in certain lights, had a beautiful pearly appearance, like frosted silver.   There were five or
six of these points on each ring.   The feet and
the head were a light brown, almost exactly
the color of an almond shell, and the green
V-shaped tail was bordered with a line of
darker brown.

He was given to me one afternoon in August, just as I was about to go out for a walk.
After admiring him, and noticing carefully his
colors and peculiar shape, I said, " I will
sketch him on my return."   But there are
some things which do not wait upon our
leisure, and a caterpillar, just ready to retire to
private life, is one.   So, when I returned to
him, two hours after, the only way he could be
sketched was with his head and three or four
front rings peering out from a well-begun
cocoon.   He had already attached the leaf (it
was a maple, as he was found near a maple
tree) to the side of the glass box, and drawn it
about him partially, and was working very
busily.

My disappointment in his special hurry was relieved, however, by finding, a few days later, and in quite a different locality, another caterpillar of the same kind, which is now before me, clinging to a spray of oak leaves, eating and resting as he chooses, with a sort of elegant leisure. Turning away from a maple leaf, he shows his preference for the oak ; clasping the stem of the leaves firmly with his ten false feet, he moves his brown head silently back and forth, while the leaf melts away before him very steadily. He has the same disposition manifested by the Polyphemus moth, which he anticipates. He never jerks about, when disturbed, or shows the slightest irritation, as do many of the caterpillars, and is so quiet in every movement that you feel sure he is well contented with life as he finds it, with no regrets for the past or speculations about the future. A perfect contrast to him is the little, jerky, impatient caterpillar of the quince, in a box beside him, who, if touched the most lightly, will actually spring up and throw himself entirely over, in the most astonishing manner. Between these extremes, every variety of disposition prevails among them. When at full length, this Polyphemus caterpillar is about three inches long ; but when hunched up like a half-closed Chinese lantern —as he now lies, eating his oak leaf—he seems but little over an inch in length.

The edge of the first ring, which comes close round like a hood over the brown head, is light lemon yellow, and the upper or second joint of the true feet, and a narrow border above the brown feet, are also yellow. The diagonal side stripes are yellow, also ; the spiracles—forming a dash near the centre of each diagonal line—are a lake-colored brown. Each one of the diagonal lines is finished at either end with a round orange or gold-colored knob (like the old-fashioned "frog button"), with a single white bristle in each.

This marvellous detail of finish in even the smallest insect excites our constant wonder and admiration.

The cocoon spun so suddenly by the first of these two caterpillars is exactly like the one cut from the wild raspberry, except that the color is a lighter yellow. The leaves are drawn over it in the same manner, and firmly glued to the cocoon. The mystery which this had seemed before was solved by witnessing him make the cocoon, just as you would better understand the Chinese ball within a ball after seeing one cut. He first bent the leaf in the position required, drawing it up at the end, and lapping it over at the side. Then he spun the fine, creamy threads of silk, weaving back and forth very dexterously, connecting the opening of the leaves with the side of the box. Contracting his body more than one half within this leafy outline, he worked himself adroitly into positions to form its symmetrical outline. I watched his work until very late in the evening, and the next morning further watching was useless. He had "wrapped the drapery of his couch about him, and lain down to pleasant dreams."

More than six months he slept in his cocoon; and now in April, 1878, he is a handsome Polyphemus moth. Very curiously, he came out just *one day later* than the one last year from the wild raspberry. That was on April 19th, and this came out April 20th. This moth is not quite so bright as the male one, and the antennæ are not so large and plume-like; but otherwise it is equally handsome. The second of the two caterpillars, as it spun up a little later, is not yet out, but the *cocoon* has been peered into, and the chrysalis, in the increasing clearness of its rings, and its active movements when disturbed, gives promise of an early exit. There is no danger of injuring

the moth by carefully opening the cocoon
which holds the chrysalis, and then its change
can be watched as it turns from a dark brown
to a lighter shade, and becomes almost transparent before it opens. Since writing the
above, a friend sent me from another State, a
box with a note—which was read before opening the box—which said,. two handsome caterpillars would be found in the box. On trying
to remove the lid, I found something was the
matter ; when lo, instead of what was promised
me, two large, scarcely completed cocoons !
My disappointment would have been greater
had I not known them at once as belonging
to the Polyphemus moth. They were busy
travellers, building as they went, and in one
short journey completing a house, with a
speed and perfection of finish which puts
greater architects to shame.

The Polyphemus caterpillar is more easily
raised than that of any of the other large moths.
The eggs are flat and biscuit-shaped, of a
chocolate color, appearing like little frosted
cakes. I have had no difficulty in rearing them
from the egg. As soon as they leave it they
are ready for the oak or maple leaf, and eat
quietly and almost continuously, making their
changes with no trouble, such as the Cecropia

and other horned or knobbed caterpillars
have. This year, 1890, I have had two come
out early in March ; the first, with broad antennæ, appearing on March 9th, and the other,
with narrow antennæ (the female moth), on
March 19th. It has remained almost perfectly
quiet, has taken no food, being, so far as I can
ascertain, tongueless, and has laid 137 eggs
on the sides of the glass box, hardly seeming
to feel itself a prisoner. The wings are not in
the least marred by flying about in the box although eight days have passed since it left the
chrysalis. The beauty of this moth is only
excelled by the gentleness of its disposition,
which cannot fail to make it a favorite with all
who prefer quiet manners to bustle and vain
show.

DOUBLE DOORS.

In Saunders' " Insects Injurious to Fruits,"
p. 175, he says of the Polyphemus moth,
" An Ichneumon fly, *Ophion macrurum,* the
same as that which preys on the Cecropia
moth, is a special and dangerous foe."

I have now (April 12, '90) a large Ichneumon fly which to-day came out of its *round*
" front door " from a fine Polyphemus cocoon.
This is much larger than the *Ophion macrurum,* and answers to the description and figure

of *Ophion bilineatus* (Say,) figured on p. 175
of Saunders.   An Ichneumon, answering to
O. macrurum, keeps him company under
an adjoining glass, and *he* walked out of the
chrysalis of an Asterias butterfly a few days
before, and is figured and described in this
volume in the chapter "Two Front Doors,"
etc.   So they are not wholly confined to one
variety of moths or butterflies, satisfied with
stolen winter quarters and food, wherever they
can obtain it.   The *Ophion bilineatus* is wholly
a russet brown in color, except his very large
black eyes (which, appearing to be six in number, two very large, very black, and very prominent, and four smaller ones, form no small
part of his head).   The two pairs of wings
are transparent, the legs long and spined, the
body very curiously curved and broadening to
the end, and the jointed antennæ nearly
two inches in length, and quite as long as the
body.   He is fond of sweets, and uses his very
curious mouth dexterously enough in securing
grain after grain of the sugar placed for him.
It is sad to look at the large well-formed
cocoon, with its usual ornamentation of the
maple leaf drawn so nicely about it, and think
the poor spinner was working so faithfully for
his direst enemy instead of securing a safe

resting-place for himself, where he should sleep into his own rightful robe of beauty. The little " front door " revealed the fact that the Polyphemus had become the prey of the Ichneumon—(the round door not being quite as large as a " shot " ;)—and on cutting open the cocoon I learned that the *chrysalis within*, instead of being eaten and broken, was scarcely marred at all ; unbroken save the small place of exit, as, in this case, *two doors* were needed for his escape. But the weight ! Instead of the solid body of the true chrysalis only a perfect *shell* remains.

VI.

HOW I CAUGHT A BEAR.

I WAS walking quite alone, when a slight noise attracted my attention. I looked about me, when, close at hand, and deliberately advancing toward me, I saw—a bear (Fig. 26).

I was not in the least alarmed, which proves how much there is in a *name*, for I did not then know he was a bear.

Determined to capture him, I armed myself with a small twig and a very small cage in the shape of a tumbler.

Instead of resisting, he coiled up quickly into a ball, was tipped into the cage, and this soon inverted over a piece of white paper on a book.

Thinking a leaf might attract him, I put a bit of cabbage leaf under the glass, and soon he was forgetful of his imprisonment in satisfying what proved to be an almost insatiable appetite.

44

He spent his time for some days in devouring leaves and taking exercise by rapidly travelling about his small prison.

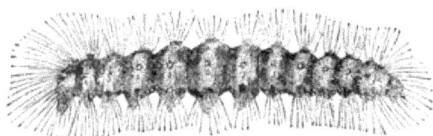

FIG. 26. YELLOW-BEAR CATERPILLAR.

Eat, march, eat, march, was his programme, until, not satisfied with one den, (Fig. 27) he made himself another, and having sealed himself in, I saw him as he was, no more. I afterward found that inside of the second den he formed another (Fig. 28). His winter quarters were secure.

FIG. 27. COCOON.   FIG. 28. CHRYSALIS.

This was in September. He slept undisturbed until March, and then he began to go about again quite freely, but in a new coat. He ate, too, but very delicately. Not leaves, but a dainty sip of honeyed sweets. In September he was a yellow-bear caterpillar. In March he was an ermine moth (Fig. 29).

A white miller, we should say, but when we part his wings we see his body is yellow striped lengthwise, and alternating with each stripe has a row of black dots. And on his wings there is the merest point of a black dot (one on each fore wing, and two on the hinder ones), so very small that you would not at first notice them. But they belong to him, and are always there. For he is not the only bear we have watched through this change, and four or five quiet, dreamy, pointed, black-dotted moths are now in a box close by me, all alike, except a little different in size.

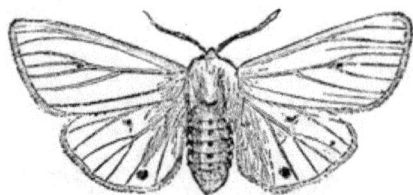

FIG. 29. VIRGINIA ERMINE MOTH.

These are the Virginia ermine moths.

In the same box are some many-spotted ermine moths, something like leopard moths; but whether tiger, bear, or leopard, the name is not derived from the nature, as all are quite meek, and much more like a lamb.

There is one of these white millers beside me now as I write. The same tiny speck on each fore wing, the same two dots on the hinder wing. He, too, went into his den in September, and came out in March (1879), so white and furry about the head that if as a caterpillar he should be called a yellow bear, as a moth I should call him a polar bear.

The golden eggs of the Virginia ermine moth turn a sage-green color (almost golden green) just before hatching, and the little caterpillars (about one twelfth of an inch in length) are lemon yellow, with dark sage-green heads.

A good deal has been said about the impossibility of raising moths and butterflies indoors. The chrysalids, we are told, should be left out-of-doors in some damp place, only secured from the worst weather, and shielded from positive storms. It is pleasant to have such proof that this is an error, as I have had the good fortune to secure from the moths themselves, who, in spite of these assertions, have opened their various prison doors for me in the past and present month (February and March, 1890), by scores. Every chrysalis of the Io Saturnia (twenty in all) has given up a perfect moth, and several other kinds have also had their opening ; among them the Chœrocampa, the cabbage butterfly, and two Virginia ermine moths. Many more chrysalids, large and small, await their coming winged life, without the shadow of a disappointment, if you judge by clearly *alive* chrysalides. And instead of an out-door, all-weather exposure, they have been in a comfortably warm room devoted wholly to them, with no extra moisture, and making no trouble. The gentle coming of these ermine moths— one day a dark-brown casket, the next, without noise or observation, a snowy-winged silent thing of beauty, the most touching thing about them always being the little tiny speck of a black dot—" one on each fore wing, and two on the hinder ones "—that these minute dots, belonging exclusively to these unobtrusive little white moths, should be given them, year after year, never varying, and so marking them as veritable "Virginia ermines," shows as much a superior care as the noting of the "sparrow's fall," or the "numbering of the hairs of our head." As silently as they come, so silently do they live their little life, sipping the sweets offered them with a delicate amber tongue, laying their eggs, small, round as tiny marbles, of a golden-yellow hue ; scarcely lifting their feathery wings to fly from one offered flower to another, and then, not waiting for their life to be *taken* from them, falling asleep *un*jarred, but not unmourned, in their little box prison.

VII.

C RUMPLE-WING (Fig. 30) came out of his winter's sleep in March.   He went in in September.   He was a salt-marsh caterpillar

FIG. 30.

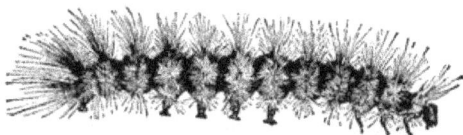

FIG. 31.   SALT-MARSH CATERPILLAR

(Fig. 31, the *Arctia acrea*).   But he seemed very much at home in an inland garden.   He was on the croquet ground, plodding his way among rolling balls and quick footsteps, when he was made a prisoner.

He lived on grass, plantain and other leaves, until he wove his yellowish-brown hairy cocoon under his glass tumbler.

I don't know why he came out of his long rest with a crumpled wing. I think he had plenty of room under his glass, and no one touched him before he was perfectly free and walking about in his queer one-sided manner. When a Danais butterfly, on coming out of his chrysalis last summer, exhibited a marred and crumpled wing, I knew it was because he had been confined in too small a space for his wings to expand fully; and the form of the pupa itself had been compressed by the position in which it was formed, so as to resemble in shape half an acorn-cup rather than a whole acorn, which it looks a good deal like when perfect. Another Danais had its wing marred by touching it very gently with a pencil's point, in the eagerness to see it expand more quickly. The slightest touch at that time will injure this delicate fabric, than which nothing in nature seems more susceptible of harm. But there was, no doubt, a hidden reason for Crumple-wing's misfortune, at whatever time it occurred. His right wings are

perfect and quite handsome. The hinder left
wing but half unrolled, and much shrivelled.
The hinder wings are a rich ochre-yellow ;
the front pair white, dotted with black and
ochre-lined. His back is ochre-yellow, with
seven black spots down its centre ; six on the
yellow, and one on the last ring of the body,
which is white. Two rows of black spots
ornament the sides, and there is one on the
under side of the body also. His antennæ
are long and graceful, and the microscope
shows them to be variegated in color, and
with spiky hairs, instead of being feathered.
His head and neck-cape are tinged with ochre.
At first he appeared so indifferent to food that
it seemed doubtful whether he *had* a tongue ;
and after being tempted in vain with sugared
water, he was left some days to work out the
question without it. But when next offered a
chance to break his fast, it was amusing to see
how eagerly he thrust out his short, amber-
colored tongue and drew up the sweets, as
a child would sip lemonade with a straw.
After his long fast, before eating, he had
strength enough to tow another moth and
two empty cocoons (which chanced to be
caught together near him) all about his box,
having entangled the claw of his foot in the

loose hair of the empty chrysalis cover.   One
or two dead moths were placed purposely near
him.   He walked slowly about them, looking
at them with the appearance of an anxious
doctor or surgeon, studying the case for a
time, and then walking off, evidently satisfied
that hope was gone when no sign of life could
be perceived.   It never seemed to occur to
him to attend to his own case, which was,
however, well enough, as it would have required as much skill to unroll his shrivelled
wing into symmetry as to put into their dead
forms a new life.   Just as he stands now, with
his head and left wing hidden under a leaf of
the blooming hepatica, you would never think
of calling him Crumple-wing.   His best foot
is foremost.   He is a fine-looking Acrea.

FIG. 32.   ARCTIA ACREA.

VIII.

UNDER THE CAPE.

THE very day Crumple-wing gave up trying to inspect others, or hold on to his own life any longer, another *Arctia acrea* came out. His brown cocoon was larger than Crumple-wing's. In fact so much larger than any one of the kind I had been watching, that a very fine specimen was looked for from it. As other Acreas had appeared, who went in about the same time, he was daily expected, and a hope (which rather grows less as moths increase in number), was indulged that his *exit* might be witnessed. A slight appearance of a disturbance at one end of the cocoon had been noticed, and he was closely watched. Just as the tea bell rang another look was given to his glass box ; when lo ! there was a small oval opening in one end of the cocoon, and the moth was rapidly advancing up the side of the box to the top. But worse than

Crumple-wing! Except that he was symmetrical, his yellow black-dotted body was
only partially covered by a very short white
cape, and two pairs of very short wings, looking like the old-fashioned double-cloak capes,
without the cloak (Fig. 33).

Watching him for a little, with a curious
mixture of wonder and pity, we left him;
when lo! on returning in half an
hour he was all right—as perfect
and handsome a specimen of the
white-winged Acrea as could be
found (Fig. 34). His cloak had
only been packed under his cape.
And this is the way he looked before he shook it out.

FIG. 33.

If another caped moth is seen before he has
shaken out his entire garment, something
more than a tea bell will be needed to prevent
a careful watching of the progress. There
was nothing of the limp appearance of a new
butterfly, to suggest any further development
of wings as necessary. His *cape* was snowy
and full and downy, and he walked off with
the buoyancy and strength of a fully developed
and perfectly dressed creature. The black
dots upon his wings are more exactly symmetrical than in any of this kind before

noticed. By actual count almost precisely
equal in number, as well as alike in shape and
size. The color under the throat is a rich
orange, and also of the thighs ; the legs being
five-jointed, alternating in black and white.
The joints resemble the divisions in the stems
of rushes, as is the case with those of most
moths when examined with the microscope.
The last joint terminates in a sharp, black
claw, with which he can cling with a force not
to be overcome without danger of breaking.
His antennæ are spiked,instead of feathered ;
and if Crumple-wing is an Arctia, as we have
supposed, and he seems to answer the description of that moth exactly, this is one of the
same class, without the ochre-lined front and
the ochre hinder wings. When at rest his
wings are roofed or sloped downward, covering
the yellow spotted body entirely.

FIG. 34. ARCTIA ACREA.

IX.

THE ARCTIAN AND ICHNEUMON.

THERE were still two chrysalids of the Arctian left, and two days after the one had stolen out from under his double cape (all moths and butterflies have the double-cape appearance), one of these chrysalids was seen slowly ascending the glass prison wall, piloted by the head and fore-legs of an ash-colored moth, creeping slowly along with his heavy brown house on his back!

It was another Arctia, or "false ermine moth," as those of this gray color are sometimes called. After a little while the chrysalis fell, and the moth was free; but, as he had "jarred in the gate" (from not being able from some reason to throw off the chrysalis so soon as he ought), his wings were somewhat cramped, and he looked like a second cousin to Crumple-wing.

After a supper of sweetened water, and upon the lighting of the gas (which always puts fresh life into every fibre of a moth), he shook out his wings very respectably, and showed his appreciation of light as the first object in life. He was of a soft glossy ash color, and his body had three rows of black dots running lengthwise down the centre and sides.

It is no slander to say that he was double-tongued, which, however much to be deprecated in human beings, is really nothing against one who uses his tongue only to gather sweets.

While some of the larger moths seem to have no tongue, the Arctians are usually supplied with two. They are coiled up side by side, sometimes joined together lengthwise, and sometimes quite separate.

The last remaining chrysalis was just like the one of the ash-colored moth, but when it opened, instead of the expected Arctian, out came a large slender-bodied Ichneumon fly! his head bright yellow and his legs alternating with honey-yellow and black. His wings are a brilliant steel blue. He resembles the Ichneumon that came out of the "round" front door of the Asterias, but is larger, and has a sword-shaped borer nearly half an inch in length, giving him rather a formidable appearance, as he comes buzzing in his "April fool!" with a bold whirr, instead of stealing in softly with the meekness of the feather-winged Arctian.

X.

THE WHITE ERMINE MOTH.

I FOUND him one November day,
  A stiffened circlet at my feet,
And made him prisoner in my room,—
  His brown coat glistening with the sleet.

Awhile he lay as still and stiff,
  As though his little life were o'er,
Then yielding to the new-found warmth,
  Shook off the icy pearls he wore,

Surveyed awhile his crystal walls,
  Shut in from liberty and—cold ;
Then built an inner prison wall,
  Closely his body to enfold.

He seemed to sleep an endless sleep,
  Silent and still so long he lay,
When lo ! in robes of snowy white
  He sprang to life one winter's day !

XI.

A HUNDRED TO ONE.

WE had been looking in vain for caterpillars on grape-vine, walnut, and sycamore, when we stopped before a large woodbine, which threw its clusters over the side of my friend's piazza, in Pittsfield, Mass. We sent our eyes upon a voyage of discovery, and peering among the thick matted mass of green—

"Oh, here is a fine fellow," exclaimed Teddy, the eager little boy being the first to discover a pale green caterpillar, so nearly the color of the vine that the similarity was his greatest protection.

"Here is another, and another! They seem to be out in force to-day; but these are so high up—how shall I reach them?"

"I 'll get a step ladder," said Teddy; and disappearing behind the corner of the piazza, he soon came back tugging the heavy steps,

and placed them under the woodbine. Now
for some tumblers. They were soon brought,
and the caterpillars imprisoned before they
knew it, eating on the leaf which had been
clipped from the vine without even disturbing
their dinner. It was well we secured as many
as we did, or even one moth might not have
repaid us ; for the caterpillar of the woodbine,
in common with many others, has a secret

FIG. 35. CHŒROCAMPA PAMPINATRIX.

little enemy, from which he is not apt to
escape. These nice-looking ones with such
good appetites, however, did not seem to have
any lurking danger. But one can not always
tell. Damocles was not the only one over
whose head hung a sword while he was enjoying his repast. Teddy selected two of the
best—not to keep himself—but for the friend
who was helping him hunt them. The cater-

pillars were soon separated ; Teddy's remaining where they were found, and the two others
going a long journey. Pretty soon some
strange things appeared on Teddy's caterpillar. He ate on, but looked rather dispirited,
as if he had caught a glimpse of the hair by
which the fatal sword was suspended. Soon he
was walking about with something all over his
back, which made him look as if he had taken
a bath, and then rolled about in a box of rice !
(Fig. 36.) The microscope showed these rice
grains to be perfect cocoons, white and silky,
and each looking as if a little cover were fitted
to one end. Something
moves inside of these.
Some of the little in-
truders are still working
on the inside of their

FIG. 36. CATERPILLAR WITH ICHNEUMON CHRYSALIDS.

rice-houses, polishing the ceiling and giving
the finishing touch to the walls.

By and by they are completed, and then
the woodbine caterpillar begins to grow
weaker. After a week or two, these little
covers begin to fly open, and as they lie back
on their hinges, out of each one creeps a small
fly, and begins to go up the glass.

He is a prisoner, and we can study him.
He is one of our old friends, a species of

Ichneumon, with ugly mouth, jointed antennæ, hooked feet, amber legs, and thin, narrow wings. He is very small,—but there are so many! The poor caterpillar cannot stand it. A hundred to one is too much, and by the time that over one hundred of these swords have pierced his body,  he was, as Teddy's grandmother said, "very dead." Here is his likeness, which an artist took for Teddy's friend. You can only see his head, one or two wings, and one foot (Fig. 37).

FIG. 37.

FIG. 38.

But the two caterpillars which took the journey seemed to escape this trouble. They both soon went into chrysalids. One drew a leaf about him, and fastened it with a few glossy hair lines to the bottom of the glass ; the other made a hint of a cocoon, with a thin network of gauze-spun threads, and twenty days after came out a pretty moth—the fore wings olive gray, banded and shaded with olive green, and the hind wings a reddish-brick or rust color. Both pairs of wings were uniquely scalloped. The chrysalids were, first a sort of mulberry color, irregularly spotted here and there, and the one which opened, growing brown (and a very dark brown between some of the centre rings), just before coming out. The second chrysalis (Fig. 39), formed some days later, is brown and dark-ringed ; but as it is a fortnight since the moth made his appearance, he is taking it very leisurely, if he appear at all. This cater-pillar and moth answer to

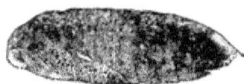

FIG. 39. CHŒROCAMPA CHRYSALIS.

the description given by Harris of the Chœro-campa, or hog caterpillar (which seems as great a misnomer as that of the Polyphemus), from a fancied resemblance of the head to that animal—the head of the caterpillar being small, and the fourth and fifth rings very large, and tapering to the small head.

The moth has been named Pampinatrix,

from its living on the shoots of the vine.
The caterpillar lives upon the grape, as well
as the woodbine. In Harris' description, it is
said that the moth leaves the chrysalis " in
the month of July, of the following year."
But this (as most other moths) has an opportunity of trying the world twice in the course
of a year. Some very large caterpillars—four
inches in length, and as large as one's finger
or thumb—closely resembling the Chœrocampa in shape, have since been found on
a woodbine in Pennsylvania. They were,
however, so completely covered with the
"rice-houses" (more than a hundred to one)
that they were not kept. Only, the parasites
were brushed from one into a box, and now
the "syrup cups" are opening, and a perfect
colony of Ichneumons are running up and
down the glass, wondering how they came to
be born in prison.

March 23, 1890.—A beautiful *Chœrocampa
pampinatrix* has come out of its brown, sharp-
pointed chrysalis to-day, and makes a pretty
picture, hovering over some blue periwinkles
in his glass box. But although their little
cup-throats have been filled with sweetened
water, he does not deign to uncoil his umber

tongue to take a sip. Just the front of the
three-grooved wheel is to be seen. No
doubt if he were flying "in fresh fields
and pastures gay" he would soon find a
use for it, but he is far too early for such
a feast and would soon die if given his
liberty. So he must use the periwinkle cup
or starve.

Close beside him, on the same box of earth,
is his exact mate, who travelled with him over
the mountains of Western Pennsylvania last
August, and who, doubtless, will not be far
behind him in the spring opening. They
were taken from an evergreen honeysuckle
and sent me as caterpillars, making their
chrysalids soon after their long journey, without a hint of a cocoon, although the Chœrocampa usually makes a very thin veil-like
covering for the chrysalis.

As described, at the time they were received
(August 15, 1889), they answered the description of those given in "A Hundred to One,"
and the moth is the same as there described.
On reading a description of this caterpillar in
Harris, and in Professor Lintner's "Fifth Report on Injurious Insects," I was at first surprised at the difference, until I remembered
how many of the caterpillars that I have raised

6

from the egg change in their appearance
almost wholly. This one I had only seen
when nearly ready for its change. At one
time it is gayly marked down the back with
spots of yellow, edged, in part, with rose-red.
And, in his description, Professor Lintner says
that shortly before changing to the chrysalis
" the color changes to a dull rose throughout."
If these assumed that color it was either a
very " dull " rose, or I was not fortunate in
the time of watching them. In some lights,
with a stretch of imagination, the faintest
hint of a pink flush may have relieved the
yellowish-brown. However this may be, the
moth is the same, so there is no doubt of his
being a true *Chœrocampa pampinatrix.* He
is a handsome moth in shape and shading,
the upper wings crossed in bands of gray and
olive-green 'and edged with a red rust color ;
the under wings being like this narrow border,
a rust-red. He is very still most of the time,
but when he does use his wings they quiver
and thrill and shake so fast that it is almost
impossible to see them. The Chœrocampas
which I have had have been those only from
the honeysuckle or kindred vines, but he is
sadly complained of as a grape-robber, eating
the leaves to the destruction of many a vine,

and cutting off the young stems of the clusters, which he does *not* want for his own use,
until the ground is strewn under the vines
with tiny green grapes.

XII.

THE UNFINISHED LIFE OF QUAKER GRAY.

I HAD a little Quaker, dressed
　In starry robe of gray,
With silken tufts of black and white
Completing his array.

His home was on a Quaker leaf,
　A poplar, silver-lined ;
On this he lived, from this he ate,
　Beneath my glass confined.

If frightened, he would drop the fringe
　Of tufted black and white,
Putting his jetty, varnished head
　Completely out of sight.

One day, when he grew very tired—
　Tired of his poplar leaf,
Tired of his small glass prison and
　His little life, so brief,

He climbed his crystal wall, and wove,
　In silence all the day,
A Quaker hammock for himself,
　Of tissue silvery gray ;

Wove it about his bead-like head,
  About his feet, so queer—
Ten feet behind, like amber spools,
  So yellow and so clear,

And six in front, like tiny horns—
  So, fastened in his net,
Day after day, as still as death,
  Hung the poor Quaker pet.

One morning, slowly out he crept,
  And a fresh suit he wore,
But, to my disappointment, just
  Like what he had before.

Perhaps a little longer waved
  His tufts of black and white,
Perhaps a little glossier grew
  His silvery coat, so bright.

Weeks passed ; a closer net he wove,
  Again of sober gray,
And, self-immured, profoundly slept
  His second life away.

More than a year for coming wings
  I watched that tight-locked cell.
Still closed remains his prison door,
  And now I know full well
That this short tale of Quaker Gray
  Is all that I can tell !

FIG. 40. COCOON OF CECROPIA MOTH, CONTAINING CHRYSALIS.

XIII.

AN EARLY CECROPIAN.

TWO rough brown oval cocoons, spun (with one flat surface fastened lengthwise to a branch) by the large green caterpillar of the *Attacus cecropia* moth, were brought in, and lying side by side, look- ed as nearly alike as possible. From one of them, on March 1st, as if to show his appreciation of spring, the fine

FIG. 41. CHRYSALIS OF CECRO- PIA MOTH.

Cecropian stole out which is now in the glass before me. The other cocoon, from eagerness to see what promise it gave of a mate, was care- fully cut at one end ; when lo, an empty chry- salis within ! Even with a microscope no place of exit was to be discerned. But his cast-off

dress was in the tomb, and it was evident he
had, with more skill and silence than the vanish-
ing Arab, gone off *without* his "tent," to enjoy
the freedom he could not have had, had he
been born in prison. I could easily believe the
remark of Harris, as I searched in vain for the
"front door," that the threads of the cocoon
of this moth "converge again by their own
elasticity, so as almost entirely to close the
opening after the insect has escaped." In
fact, I could omit the "almost." The change
is indeed marvellous from the large light-
green and coral-dotted caterpillar (making one
think of a cactus stem that had concluded to
walk off), to the gray, white, and cinnamon-
brown moth (Fig. 42). The six legs and most
of the body are cinnamon-red. The broad brown
antennæ, with central amber stem, come out
from the front of the rather small cinnamon-
colored head. Just back of this a neat white
collar, and then the tufty brown extends back
half an inch, and from it proceed the wings.
Then comes a narrow band of lead color, and
the rest of the body is ringed with black,
white, and cinnamon-red, alternating. Along
each side are seven round cinnamon-red spots,
bordered with white. The finish of the hinder
wings, in heavy lines of alternate gray and

black, reminds one of a pheasant's wings ; but
above this border is a line of the red, and
above that a narrow line of white.   In the
rich furry grayish-brown of the hind wings
are two large crescents of red and white. The
front wings have no white in the stripe above
the beautiful scolloped gray and reddish-white.
They have an eye spot near the edge, of very
dark brown or black, edged with white.   It is
a very rich moth, though not as soft in the
harmony of its colors as the Polyphemus.
Like that it is very gentle in its manner, keeping almost entirely quiet during the day, and
flying but little in the evening.   Its eyes are
black.   If it has any tongue it is not to be
seen, at least while the moth is living, even
with a microscope ; nor can the moth be
tempted to use it.   Its wonderful tenacity of
life, when this fact is considered, is very remarkable.   It will live about three weeks
apparently without food, and pays slight attention to any thing ordinarily used in putting
moths to sleep ! The moth stands most of the
day with its wings almost together, but will
slowly open them to their full extent if blown
upon slightly.   The caterpillar may be found
upon the apple, cherry, or plum-tree, and
changes, from being at first a deep yellow,

FIG. 42. THE ATTACUS CECROPIA MOTH.

to its last coat of handsome light green, before going into its chrysalis (Fig. 41). It is said by Harris to come out in June, but, whether on account of the very mild winter and the usual difference of climate between Massachusetts and Pennsylvania, or as a surprise to insect lovers, this Cecropia is three months in advance of that season. So early an exit will make less difference to a moth without a tongue. There are three fine cocoons of the *Attacus cecropia* before me, in a box (opened at one side so that the chrysalids can be watched), as I write (March 29, 1879), and by the transparent lines between the rings one of them shows it will soon release its impatient prisoner. The Cecropia worm spins its cocoon invariably alongside a twig or branch, as shown in the cut, when in the orchard or wood. But one of these three (the caterpillar of which was confined in a glass jar) made his cocoon of the usual shape and texture, except that the material is a richer, glossier brown, but it is not attached to a stem. It was fastened to the side of the glass by a heavy web of dark silk, very much darker than the cocoon itself, which is a handsome russet-brown. The inner lining is very glossy, and the whole fully three inches long.

The Cecropia moth is more difficult to raise "in prison" than either of the other three large moths (the Luna, Polyphemus, or Prometheus) of this genus, Attacus.

On July 8, 1889, I received a box with a large number of the eggs from a friend in New Jersey, all of which came out (several having hatched by the way). One peculiarity I noticed at the start with them all—they do not eat the egg upon leaving it, but just enough to allow their escape. They began at once to eat the lilac,—and the pear,—as well as the currant-leaves which were given them. I have been more interested in this fact (of which I made a note at the time), from a discussion of this point between two of our best entomologists, since my notice of the Cecropias, one asserting that the eggs of butterflies—and I suppose moths as well—were *always* immediately eaten, and the other as strongly saying it was *not* always the case. I have since watched many different caterpillars in this respect. The Danais eggs were eaten usually as soon as vacated,—every vestige gone in a little time, except in a *few* cases, where the Asclepias leaf was ready at the little open door, and probably had a fresher attraction for the escaped prisoner than his prison walls. But in other cases, notably in nearly a score of eggs of the rosy Dryocampa, the empty shells remained, and are preserved in an insect-egg collection, whole enough (and brilliantly glossy) to show the shape and material perfectly. This is also the case with a group of most exquisite pearly eggs, found opened and deserted, of the still unknown occupants. So that unless an entomologist knows *every insect* of the millions, in their first and latest habits, it would probably be safer not to make assertions *for the whole*, which *a part* may rise up and prove mistaken.

The Cecropias, in their *first* stage, are chiefly black, and spined on all the rings, each spine (plainly to be seen with a microscope) having three or four hairs, or finer spines. A *few* were yellow in this stage (as Harris gives them), but most of them were quite black. In the *second* stage, the black coat is exchanged for one of russet yellow, with black spines, which are each spined, in wheel-forms, with one russet-yellow spot, or knob, on each cluster of spines. In the *third* stage they are *bright* yellow, the wheeled spines jet black, like *spokes* of spun black sealing-wax, from *hubs* of clear garnet beads, one bead, or knob, being the centre for four, five, or six spokes.

In this stage also, the *second* and *third* rings
are very handsome. On the top of the back,
on each of these two rings, are two *crimson*
" hubs," spoked with black. The next to the
last pair of knobs, or hubs, are a pale indigo
blue, and this color is hinted at in several of
the knobs toward the end, all being spined
with the jet black spines. In the *fourth*
stage the color is a very pretty light-green.
There are large coral-red warts on the second
ring, and smaller ones of the same color on
the third, while on all the others to the
eleventh there are yellow, egg-shaped promi-
nences, beside which there are two rows of
light-blue beads, or warts, each side all the
length, and one row of the same color on the
side (below these) of the first five rings,
giving the whole caterpillar a " coat of many
colors," sufficient to excite the envy of all
his acquaintances. But he pays dear for that
part of his ornamentation which consists of
raised work, and which not unfrequently costs
him his life when attempting a change of
garment.

[NOTE.—The easiest way to transfer the imago from
the box where it has completed its change to the cyanide jar (and which does not necessitate touching or
alarming the moth), is by holding a narrow-folded strip

of paper before it, upon which it will invariably step, and both paper and moth can be dropped quietly into the jar.   So many moths are rubbed and defaced by rudely taking them between the fingers and thrusting them into the jar that it seems to me this simple and successful way is worth mentioning.]

XIV.

I HAVE been April-fooled several times within the last hour. Not by a person ; but by a moth—my beautiful rosy Dryocampa. It was no April fool, but a pleasant surprise, its coming out this April morning after its long sound sleep, never once moving, in the black ring-notched chrysalis, since it went into it on the twenty-sixth of last August. A beauful little creature it is, especially t h e under wings, which

FIG. 43. DRYOCAMPA RUBICUNDA.

look, more than any thing else, like a stray rose-tinted sea-shell, such as one sometimes finds, nearly transparent, and almost as flat as a rose petal.

I was trying to sketch it, and it would stay so perfectly still that I would think, " Now, I

79

shall have a good chance!" and lo! when one wing, or the crested head, was half drawn, away it would fly. Recaptured, I would begin again, and with the same success as before ; so that when I had about six half-finished sketches, in as many different positions, I remembered it was the first of April, and quietly put it under glass, until the picture was secured.

There are but two colors, rose and yellow. The upper wings deeply bordered with rose behind, and broad epaulettes of the same color. The under body and feet are rose color also, and there is the faintest hint of rose on the under wings, which are studiously kept out of sight. All the rest is a bright yellow. The head is tufted, and the eyes are set so far under in front as not to show, unless you peep under the tuft, where you see them, black and round, close to his little front feet. There is a triangle of yellow, bordered with red, between them, and a little triangular tuft of the same color at the base of each of the delicate antennæ. Much of the time when the moth is at rest these antennæ are completely hidden, by lying back close along the edge of the front wings (like those of the *Quinque maculatum*), so that you would be apt at first to think he had

none. They have about thirty joints, as near
as one can count them when in such constant
vibration as they are pretty sure to be when in
sight. He will keep perfectly still two hours
at a time (if you are not attempting to take
his picture), then fly about wildly for two or
three minutes, and then for hours remain
immovable, as if dead. This one prefers to
stand showing but three feet—two on one side,
and one on the other,—and no coaxing draws
out the shy foot. The under wings are kept
out of sight, except a little margin in front,
near the head, which shows a small crescent of
faint rose color below the upper wings. The
antennæ of the female moth are simple, like a
little strand of beads, while those of the male
are spined, being larger, as are those of all
male moths. The only other moth of this kind
which I have seen went into the chrysalis state
in the summer (July 5, 1877), and came out
the last of the same month (July 27th), perfect-
ing in that time the work which—however
soon completed in the fall caterpillar—remains
out of sight nearly half a year. Harris, in the
description of the rosy Dryocampa, says,
"The caterpillar is unknown to me," and I
have not seen it described elsewhere. The
two which I had (one of which I watched
6

through the change into the chrysalis) were taken from beneath the maple tree, and were nearly ready for their change. They do not spin any cocoon, nor attach themselves to the glass (like the caterpillar of the Danais and also of the Asterias, and others), but work off the caterpillar skin—the chrysalis first appearing of rather a bright green or yellowish color, and soon becoming quite black.

The *summer* chrysalis would move, when touched (advancing on the paper with a peculiar gliding motion, by means of the toothed edges of the rings) ; but the winter one was never seen to move a hair's breadth. The caterpillar has twelve rings, is a pale pea-green, and striped lengthwise (which gives it a somewhat checkered appearance) in narrow stripes of a little deeper shade of green. The head is a russet-brown color, and there are two soft black horns on the second ring about one third of an inch in length. The under side of the two rings before the last are a purplish-brown, edged all along with short, black spines. There are a few short, black spines on the last two rings, and the V-shaped tail is edged also with a border of them, as also is a line along each side of the body. There are minute black warts symmetrically arranged about each ring,

about five on each. It is curious to compare
a butterfly or caterpillar either with another or
with some written description, and notice the
exactness of repetition in spot, spine, and
marking of every sort. In writing as minute
a description of a certain caterpillar as could

FIG. 45. CHRYSALIS OF
DRYOCAMPA RUBI-
CUNDA.

FIG. 44. CATERPILLAR OF DRYOCAMPA
RUBICUNDA.

be given from counting both spots and spines,
I was pleased to find afterward a printed description answering count for count. There is
not always the same similarity in their *cocoons*,
as they will accommodate themselves to cir-

cumstances rather than give up the idea of
building their home. The Polyphemus will
always draw leaves together in a graceful manner about his cocoon ; but one, from whom I
took his supply of leaves, when about to spin,
made his cocoon without it. It is true he was
the only one of several which I had who died
in his cocoon ; whether from mortification that
he was obliged to deviate from his usual plan,
I never learned. But the *chrysalids* (except
from some malformation) seem to be as exactly similar as the moths and caterpillars.

The eggs of the rosy Dryocampa moth are
very handsome. To my surprise I found
(August 28, 1889) a beautiful cluster of these
eggs (seventeen in all) on a leaf of maple ;
some of the tiny caterpillars just emerging
from their bead-like cells. With the microscope I could at once identify them as those of
the rosy Dryocampa, which I had not before
known in its earliest stage. The eggs were
very shiny and glass-like in finish, light pea-
green, and globular, with a plain surface. The
little caterpillars, with dark, almost black, heads,
and bodies pea-green, striped lengthwise, and
the two little horns or feelers on the second
ring from the head, showed them at once to
be rosy Dryocampas. In coming out they

had only eaten the *roof* of their glossy green
houses, and the lower half of the little circles
still glisten on the maple leaf where they were
first found. The caterpillars grew well in confinement, and each one of them now lies, a
fine chrysalis, waiting for some fair day to come
out from its dark, notched case, with rose and
yellow wings, triumphant in the change.

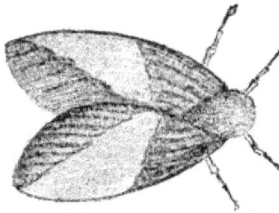

FIG. 46.   DRYOCAMPA RUBICUNDA MOTH.

XV.

THE SATURNIA IO.

THE handsome Indian yellow moth, *Saturnia Io*, was one I learned backward. Finding a beautiful moth of this kind on a fence one evening at twilight, I secured it with delight,

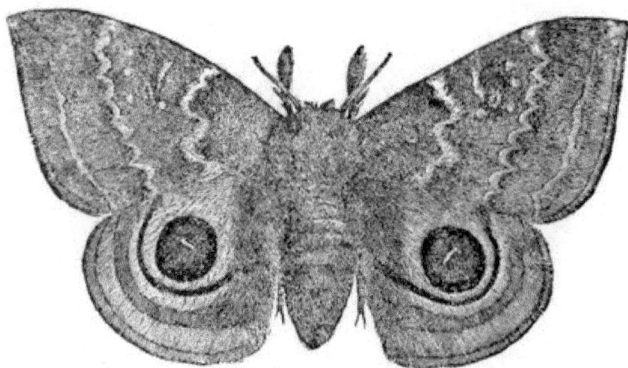

FIG. 47.   SATURNIA IO (FEMALE MOTH).

but with no knowledge of its name or from what sort of chrysalis or caterpillar it had come.   After keeping it some days, I found

it one afternoon apparently dead. Touching
it, or moving it along even, with a pencil,
betrayed no sign of life, and it was carefully
placed in a box containing several other specimens. While reading in the same room that
evening, I was startled by an unusual sound,
which, as I was alone, was a little annoying at
first, but soon I perceived the noise came from
the direction of a box of moths! And sure
enough, my *Saturnia Io*, far from being dead,
had taken occasion to call on each particular
moth in the collection in the most unceremo-
nious manner, ascertaining to its entire satisfaction, if not to mine, that none of the others
had been put away (not to say buried) alive.
Some delicate wings were detached from poor
victims unable to return this unmercifully swift
whisking about ; and before the Io could be
safely transferred to solitary confinement, he
had brought confusion out of order in the
most undesirable manner possible. So began
my acquaintance with Io. In the latter part
of the following August, a caterpillar was given
me by a friend, of a kind I had not seen before,
and soon I found two others like him. They
were between two and three inches long, and of
a light pea-green color. The twelve rings were
each starred with a cluster of green spines,

tipped with a dark purple, looking almost black. These were sharp and thorn-like. A line of purple brown ran along the lower part of each side, bordered on the lower edge with yellow. The hinder prop-feet were a dark brown; the eight middle feet purplish, with a brown finish

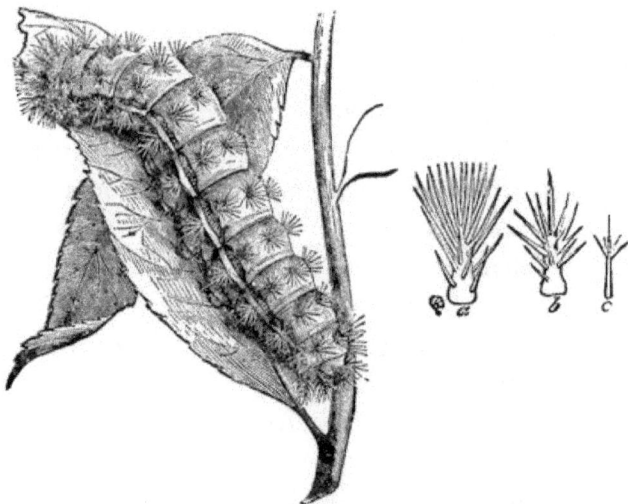

FIG. 48.   SATURNIA IO CATERPILLAR, WITH THE THREE GRADES
OF SPINES.

at the bottom. The three pairs of true feet were purple. The head was green like the body, while the mouth was purple like the feet. The first ring was so completely covered with spines as to hide his head entirely when bent forward, as they usually were. There were six

sets of these stars on each ring, except the last
two (and five on each of those), and on the
first four rings, which have on each side an
extra cluster very low down.

These spines are very stiff, and remind one
of porcupine quills. The purple-brown line
along the side, which begins at the fourth ring,
bends down to the hinder prop-feet, leaving
five clusters on the last ring. On each side
of every ring is an oblong vertical breathing
hole (spiracle), as in nearly all larvæ; for
though these differ in number and some other
respects in different caterpillars, yet their arrangement is uniformly symmetrical, and usually each segment is furnished with a pair.
Examined with a microscope, this spiracle has
first a vertical white centre line, around which
is an oval of brown, and this again bordered
by an outside oval of jet black. He looked
like a moving strip of star moss. He refused
clover, dogwood, and elm, all of which they
are said to like, probably because when taken
he was about ready to become a chrysalid.
There are in each star about thirty spines.
Three shorter ones usually in the centre, a
second circle about these three, and again a
third, which are still longer. Some of the
spines, especially in front, are not tipped with

purple, but end in delicate long hairs. While
really pretty, they are a formidable-looking
caterpillar, and the sting of the spine is said to
be as severe as that of a nettle. So curious a
caterpillar was not difficult to be found described, and I soon learned, if these went
safely through their changes, I should have the
Saturnia Io moths. In a very short time the
three had spun their cocoons and retired for a
winter's sleep. Two of them seemed to strike
up a close friendship at once. While the third
went off to a corner of the box and spun his
cocoon independently, the other two worked
closely side by side, form-
ing a twin cocoon, joined
together entirely on one
side, and looking not un-
like a double covered cra-
dle. This being a new
departure (as in the case
of the Polyphemus cocoon,
without the leaves on the
outside), only one of the pair survived the experiment !

FIG. 49.   CHRYSALIS AND
COCOON OF SATURNIA IO.

On the last day of winter (February 28,
1878), one of the covered cradles opened, and
a beautiful female moth came out (Fig. 50), just
such an one as had made the bustling expedition

among the box of specimens in the fall. On the
third of March the single cocoon opened, and a
male *Saturnia Io* appeared (Fig. 51). It is of a
deep Indian yellow, with the four wings oblique-
ly marked with purplish red, and a number of
spots on each, close together, near the middle of
the wing, which, have been thought to resemble
the letters A H, and which, with a little help
of the imagination, do look more like those

FIG. 50.   FEMALE IO.

letters than any thing else. His mate is much
darker, with less of the yellow and more of the
brown and purple. Instead of the letters A
H, there is a three-scalloped spot of rich, deep
brown, edged with gray. The head is a rich
snuff-brown, very velvety, and the handsome,
velvety feet are of the same rich color. The
other half of the double cocoon remained unopened.

After writing the above, it was my curious
good-fortune to find seven of these large caterpillars on one blade of Indian corn.   There
was not another to be found in the small field,
and how these had chanced to congregate in
such camp-meeting array was a mystery.   They
are " processionary " caterpillars, and although
I had read this, I should not have realized it
but for the curious sight which having so many
at once afforded me.   After they had been put

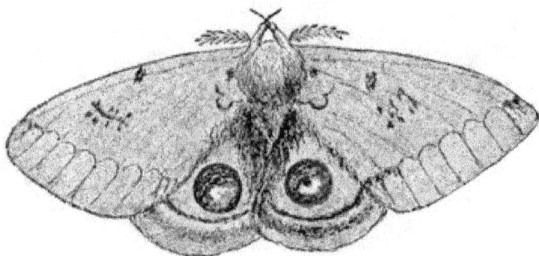

FIG. 51.   MALE IO.

under a large glass, it was a new and amusing
sight to watch them march around—one length-
ened, mossy line of green, all touching one
another and walking as fast as if quite alone.
They preferred the green leaves of the corn to
any others which they are said to like and will
eat.   One after another they made their seven
cocoons, and lay through the winter just passed,
side by side, a little hamlet of sleepers—houses

so still and apparently unoccupied as to have suggested a " Deserted Village," but whose occupants I knew were only waiting to surprise me on some coming spring morning with a regular Chestnut Street parade.

And the spring opening has come. Three of the sleepers have left their black, moveless, chrysalid homes. One has lived his little life, and two rich brown and purple ones are in a box near me (March 31, 1879). One of them has just made a pretty picture by flying upon a fresh light-green blade of Indian corn (planted in my room expressly for their pleasure), almost, but not quite, too frail, in its own forced and tender growth, to support his swinging and fluttering little body. The corn was not for them to eat, as these moths may be classed among the tongueless ones, nor could they get any good from the green blades, had they ever so long a tongue. But if it were June, and they were in the cornfield, there they would deposit their eggs for the future star-moss caterpillars—more than two dozen of which, rather large, and of a clear, golden yellow, are now in a box, with a leaf of the corn for any possible coming need.

The pupa is black throughout, so that there is no change in it to indicate the coming of the moth, as is the case in so many of the brown or other lighter-colored pupæ. The rings, however, become a little wider apart, and the spaces a little clearer, perhaps, between them. The end of the pupa, opposite the head, when looked at with a microscope, is drawn in a little curiously, reminding one of the peculiarly pretty bud of the laurel blossom.

Another cocoon has since opened (April 10, 1879), and a lemon-yellow and variegated male Io has shaken out his beautiful wings, handsomer in his light spring suit than any of the others.

XVI.

SILVER GRAY.

STANDING on the heart of a blush rose, with his richly shaded, silvery wings fluttering over its soft petals, my *Quinquemaculata* moth makes a fine picture. His wings are spread just enough to show five orange spots encircled with black, which ornament each side of his body and give him his name. But the back of his head, between the shoulders, is his chief beauty. It is a rich, soft gray, curiously and regularly watered with black and white wavy lines. Of his six legs the last two pair are branched with three delicate spines. The eyes are very large and velvety black. The antennæ are not feathered, as are those of the Polyphemus moth, but many-jointed, tubular, and finely pointed at the tip.

These antennæ are about an inch in length, and usually lie back close to the side of the

body, seeming to form a corded edge to the
upper wings, the points lying just under the
wing. You would at first say he had no
antennæ; but watch him a little, and they
will soon be very apparent. The tongue is
four or five inches in length, but when coiled,
looks like a small wheel set between two
feathery side pieces.

When freed from his chrysalis, his first care
seemed to be for this long slender tongue,
which had been so specially cared for, during
the chrysalis state, in its curious pitcher-handle
sheath. He unrolled and shook it again and
again, curling and smoothing it as a child
would a dandelion stem, and then reaching up,
touched the top of the glass box (quite a high
one) several times. Then he coiled it up
quickly, and that was the last seen of his long
tongue, except the hint of it in wheel form.
Although tempted by fragrant flowers and
sugared moss, he would not be induced to uncoil it again. A "greater green orchis," with
its immensely long nectary of sweets, would
no doubt have given him an opportunity to
satisfy his hunger in a becoming manner; but
no such flower was at hand, and scorning
to use so remarkable an organ upon any ordinary repast, he quietly became a martyr to his

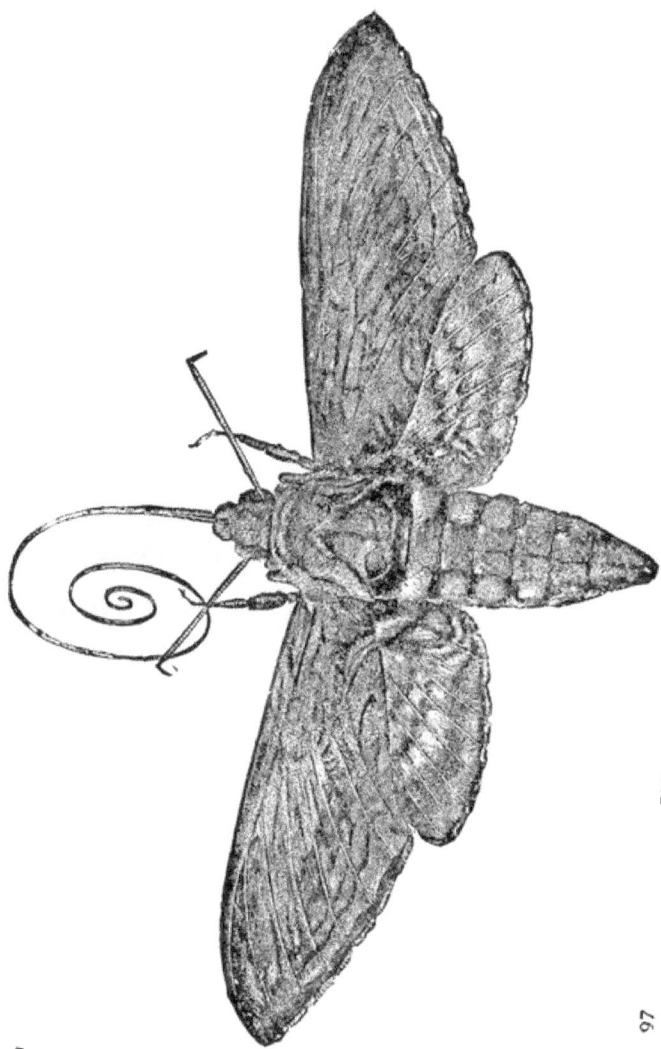

FIG. 52. MACROSILA QUINQUEMACULATA MOTH.

7

sense of propriety, and died from hunger in
the midst of plenty.   And what is this dainty
creature ; or, rather, what was he ?   You will
exclaim when I tell you he was the revolting-
looking, large, green tomato worm.

Snappish and really dangerous in that form
—requiring to be taken with great care—the
change in his disposition seems as great as
that in his external appearance.   Although
he does not equal the Polyphemus in gentle-

FIG. 53.   LARVA OF THE QUINQUEMACULATA MOTH.

ness (and I have seen no moth that does),
still he is timid and quiet; although I fancy
when touched there is a trace of the original
disposition in the short, quick flutter he gives
in response.   It has not been an easy matter to secure this moth.   It is a sphinx, and
like all this class the caterpillar buries itself in
the earth to go into the chrysalis form.   Several large specimens of the tomato worm were
caged in boxes, upon earth, and fed with to-

mato leaves. In due time they all disappeared
in the earth. The same curiosity which leads
children to take up seeds once or twice to see
if they have sprouted, led to several attempts
to see if these chrysalids were formed. Though
Nature cannot be delayed, neither will she be
hurried.

At length, all the earth being shaken from
them, two large well-formed chrysalids appeared. These were allowed to lie upon

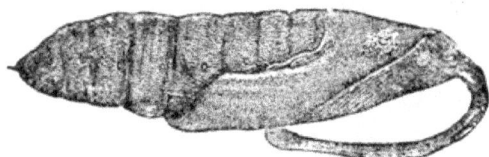

FIG. 54. CHRYSALIS OF THE QUINQUEMACULATA.

the earth all winter. They showed signs of
life until March, when they shrivelled a little,
and would no longer move when touched.
They are now "hardened cases," with no hope
of change.

This was too great a disappointment to
bear without some attempt at remedy, and the
thought was suggested of digging where last
year's tomatoes had grown, to see if any un-
watched ones had survived. The gardener
soon brought two fine chrysalids to light.

They were laid on boxes of earth in the empty glass case which the others had occupied, and Silver Gray broke the bands of one of these yesterday. The large moth made its exit at the usual place between the shoulders, leaving a mere parted line in the almost unbroken chrysalis. Even the long tongue-sheath was not broken or loosened from the breast. These two chrysalids were alike. The first two differed only in the tongue-case, one having the pitcher-handled case, as in the engraving, while the other had two short, straight cases, side by side.

What may we not believe possible in transformation, when we see the forbidding tomato worm, after a dark underground existence, come out into the silvery beauty of the *Quinquemaculata ?*

Shall *we* fear "the dark prison of a tomb," since the same power that opens the chrysalis rolls the stone from the long-sealed sepulchre ?

FIG. 55.   CERATOMIA QUADRICORNIS (HARRIS).

XVII.

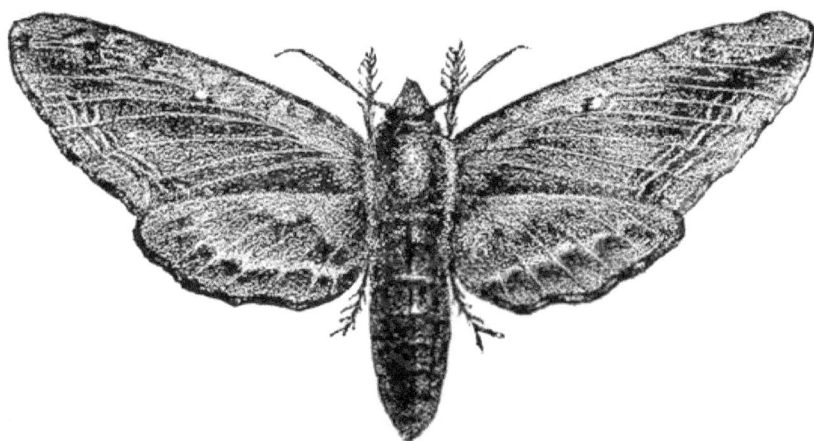

CERATOMIA QUADRICORNIS (HARRIS).

I HAVE dipped the bells of the lily of the valley in sweetened water and put them in the box where my fine moth, *Ceratomia quadricornis*, may have a rich treat.   But, as usual with these long-tongued moths, he scorns the feast, although his mouth waters for it, as one can see by the way the little brown wheel moves between the "tongue-cheeks."

For five minutes he has had his head buried in a lily bell, but not a muscle moves, as I watch him with my glass.   Had he plunged it in himself, he no doubt would be sipping sweets, but as it was put over him, in a way he did not understand (nor resist), he simply scorns a forced meal.

He is a very richly shaded moth, with, however, no bright colors. He is a little larger than the Philampelus, or "vine lover." He is a rich brown, light, with very dark wavy shadings, like watered silk, with a very little white. The body has five lines running lengthwise, of the darkest brown shade.

The caterpillar lives upon the elm. This one came to me in a box from a friend in New Jersey, August 10, 1881. The moth derives its name from the large, green, rough caterpillar, which has four horns on its shoulders (Fig. 56). These horns are evenly and curiously notched.

FIG. 56.   CATERPILLAR OF CERATOMIA QUADRICORNIS.

There are seven diagonal lines on the sides, and down the centre line of the back there is a row of notches like the teeth of a saw.

There is a horn or spine on the end of the body—a continuation of the notched line of the back. I had found one of these on an elm in Pittsfield, Mass., two years before, which died in the transformation to the chrysalis. It is very difficult for these horned caterpillars to make the change into a chrysalis. And as I only laid the one first secured on the top of a box of earth he had a very poor chance to effect it. He was, at first, a fine noble-looking fellow, but in his efforts to change into the chrysalis he became the most forlorn-looking of creatures. His little amber-colored feet were brought together (in pairs), almost touching, like folded hands. You could only see that he breathed by their gentle rising and falling ; and even this would cease for such long intervals that one would think him dead. His skin shrivelled until it looked like the brown netted meshes of a nutmeg melon rind, and after a few more faint efforts he lay still, not to move again.

When the fine specimen given me, August 10th, was first received, he was evidently ready for his change. He was placed upon a box of earth, and in less than an hour (after describing a very correct horseshoe in one voyage of discovery on top of the earth) he went quickly out of sight, and remained for months undisturbed. In March he was uncovered, a fine large brown tongue-cased chrysalis, and watched as his wings grew farther apart and a little clearer, until, fortunately when my eye happened to be upon him, I saw him break his casket, and step briskly out and walk up the side of the glass box (upon the ribbon edge) in about three seconds of time. I say fortunately, because those who watch chrysalids know how very certain they are to spring upon you in full-dress when your back is turned for a moment. Out of eleven *Saturnia Io's* which opened this spring, equally watched, *not one* was seen during the exit. It was perhaps an hour before his wings were entirely shaken out, but such perfect unfolding, without wrinkle or seam, after such long and *tight packing,* is not seen from any traveller's trunk!

FIG. 57.   PHILAMPELUS ACHEMON.

XVIII.

PHILAMPELUS ACHEMON.*

"ONLY honey-dew, and sweet manna !
No more grape leaves for him !"
Strange words to say bending over a large ungainly caterpillar, one would think, and yet I
knew why he was hurriedly making his way
out of sight in the box of earth on which I had
not an hour before placed him, and why, as
well, he had turned fròm the fresh leaf of grape
I had just brought for his supper.   No more
grape leaves—done with coarse food and low
grovelling life ; no more crawling and creeping

* Vine-loving Achemon.

and half-blind existence.   A long, quiet rest, out of all sound and sight, and then a fresh, bright awakening to soar and sip from the daintiest flower chalices, in the exquisite garb of the gentle *Philampelus achemon.*   This I knew was before him, although I had failed the year before in an attempt to see him through these changes, farther than the chrysalis, which I had probably prevented reaching its perfect state by exposing it to the sunlight before I had learned that it should have darkness rather than light until the time of its winged awakening.

It was on the first day of October ('81) that I spied him, at the close of a game of croquet we had been enjoying, slowly making his way down a fence-post, beneath the grape-vine. An odd enough, and not very prepossessing-looking fellow (Fig. 58), in his russet-brown

FIG. 58.   CATERPILLAR OF PHILAMPELUS ACHEMON.

dress with diagonal cream-colored side-stripes, six on a side (made up of a sort of chain of twisted oval spots), and a curious staring eyespot on the top of the last segment of his body. More odd still, when disturbed, he drew his head and the next three rings of his body into the fourth ring, making a monk of himself without ceremony (Fig. 59).

FIG. 59. CATERPILLAR WITH HEAD WITHDRAWN.

Placed upon a box of earth, (covered with glass), in less then half an hour he was out of sight.

Tipping the box carefully, a few days after his disappearance, letting the earth slide from him to disclose his successful change into the chrysalis (a large chestnut-brown case), I covered and put him away (Figs. 60 and 61).

There he slept until May 14, 1882, when, looking at the box again as I had for some days been doing, his hour of triumph had

come! I saw, standing above the open chrysalis the beautiful *Philampelus achemon*, his

FIG. 60.   UPPER SIDE OF CHRYSALIS.

wings trembling and expanding into his now perfect dress.   His eyes are very large, the antennæ long, slender, and pectinated, and you have no need, as in the case of the *Polyphemus*, to search for his tongue, as its large coil shows like a brown wheel between the deep rich velvety side-pieces, or tongue-cheeks, which enclose it.   As he stands now, on a bunch of cherry blossoms in his large glass house, with his curiously scalloped, or cut-in wings expanded three inches across, we can but wonder at the secret of the change which went on silently in the buried chrysalis.   The wings are a beautiful ash color,

FIG. 61.   UNDER SIDE OF CHRYSALIS.

with a faint reddish tinge ; the fore wings ornamented with two very rich dark velvety-brown spots nearly square in shape, and the hinder wings are of a bright pink, bordered behind with ash color. There are also two triangular brown spots, of the same color as those on the fore wings, on the thorax. He is a very quiet moth, resting for hours in one position, and not at all vain, as he takes no pains to show his chief beauty, the exquisitely colored hinder wings, which are almost entirely covered by the front ones. Thus far he has not been seen to uncoil his tongue, though tempted by sugared water on moss and flowers, and, last and chiefest, by a leaf-cluster from his own vines, which "give forth so sweet a smell" that if he had any reminiscences of his former life, he would, one might imagine, be induced to prove himself still entitled to the name *Philampelus.*

XIX.

THE FOX-FACED MOTH. [ADONETA SPINULOIDES].

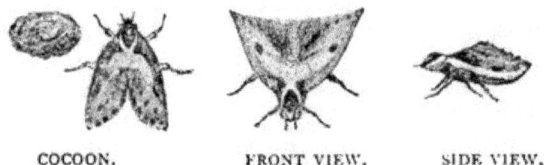

COCOON.   FRONT VIEW.   SIDE VIEW.

FIG. 62.

ONE more look at the little round, smooth chrysalis, not larger than a pea, which has been watched carefully since last August, and lo! standing meekly by its open house is the delicately fringed, bronze-shaded moth (April 14, 1882) so long waited for. It is one of the limacodes, so difficult to bring through from caterpillar to imago. Once before (December 2, 1881) one of this kind, a male, came out, but before it could be identified it was so broken, in removing to a new box, after mounting, as to be unrecognizable, so

far as determining its species was concerned.
On the 17th of August, 1880, the first caterpillars I had ever seen of this moth, except
one, the year before, which soon died, were
found on a small plum-tree in the garden (the
same from which the first was taken), and
they were found now in large numbers.
Twenty-two were secured that day, and in a
note-book of that date are simply described
thus : " They have three or four diamonds on
the back—three purple diamonds, on a yellow
ground ; the rest of the caterpillar is green."

Three days after, August 20th, is noted :
" Three of the diamond caterpillars have spun
up. The cocoons are small and hard, smooth
and parchment-like, and each is glued to a leaf
of plum. One is yet unchanged on the leaf.
These were under a tumbler. The rest were
in a glass box, which being ribbon-bound and
not perfectly tight at the side, allowed a few to
escape. Two were found and put back, so
that there were fifteen or sixteen left. They
are very handsome under a microscope. They
are pea-green and spined down the edges at
the sides. There are eleven pairs of spines,
fringed with delicate black hairs. The three
pairs in front, and the three pairs behind,
are larger than the intermediate ones. The

spines are scarlet, and each one branched
with five smaller spines, which are pea-green.
The diamonds on the back run into each other ;
three toward the head and two toward the
back, and, under the microscope, there is much
work on these diamond-shaped spots. There
are three straps across each one of the larger
diamonds, and these are buttoned at either end
of the strap.

"The space between the diamonds on the
back (which space the microscope reveals,
although to the eye alone they appear to join)
is yellow."

Again under date of August 24th : "The
fourth of the diamond-backed caterpillars spun
up. A small, round cocoon, smooth like the
rest, but pea-green instead of brown."

At length, there were more than a dozen of
these small chrysalids, but of them all only one
reached the imago state, and appeared as early
as December 2d, as mentioned above. Its
description, carefully written at the time, is
given December 3d :

"It is of a rich brown and light drab. Char-
acteristics : Large black eyes, low down in
the head ; a hairy crown-like tuft, rather square
and flat on the top of the head, which is dark
brown, edged with light drab ; legs slender and

silvery; thighs large and spined; no tongue
visible." After he was ready to mount, and so
was quiet enough, I counted the joints of the
antennæ with a microscope and found them
about thirty-six. The front of the head
seemed pointed and nose-like. On each side
of the nose, just over the eyes, are the
antennæ, amber-colored. They start from
almost the same point. This peculiar shape
of the head gives it a fox-like appearance, and
having been struck with this, I was pleased
when a friend noticed it, and remarked: "It
would be strange if he should show fox-like
habits." The legs are very slender, and in
both the specimens I have had, seemed to
come off very easily at the first joint—once
from being caught in a drop of sugared water
the leg was left in the sweet, and with no gain
to the mouth, as there was no notice taken of
food. After losing his leg (the right front
one), he would use one of the antennæ in its
place, and turn the other back to edge his left
wing!

August 22, 1881, two more of these bright
little caterpillars were found, and the next day
one more. These were on the same plum-
tree, and although several other plums were
near, no trace of one has ever been found on
8

any of them.   These three were all that could
be found last year, and in a day or two, two of
these were covered with little ichneumon rice-
cases.   So that but one cocoon was made, and
this, fortunately, survived and came out in
April; so much later than the other moth of
last year, that it had almost been given up.
The case in which the moth was enclosed,
inside of the cocoon, came out of the cocoon,
a clear skin, showing all the marks of the moth,
even the antennæ.   There were six eggs fastened upon the leaf in two exact rows—amber
colored.   These being on the under side of
the leaf, were put into the cyanide jar, unnoticed, and thus probably had their life destroyed.   Whether the egg-life will survive
that which killed the moth, will be an interesting question.   The markings of this moth are
not so distinct as of the male one, and the
body is somewhat larger.   The colors are
similar, although the contrasts in shading are
less marked.

XX

LIFE IN A BASKET.

IN a recent number of a magazine a correspondent asks, "Can any one name a caterpillar which lives on evergreen trees? It carries its cocoon on its back. The cocoons have evergreen needles hanging down the sides."

This curious caterpillar, usually called Basket-worm, from its basket-like case, belongs to the Psychadæ family. On the 5th of August, 1879, I received some of these curious baskets, from a friend in New Jersey. The baskets were bottle-shaped, rough, and covered lengthwise, with bits of arbor-vitæ. One was drawn up close at the neck like a sack—(the Germans call them sack-trager or sack-bearers) and I supposed, as it was perfectly still, that it was dead, or had changed to a chrysalis. Another at once put out its head, and the three following rings of its body, and began to walk up

the glass box (ten or twelve inches high),
drawing his basket along with him, and so
walked to the top, and across one end clinging
by the ribbon binding. The same day he
began to fasten a thread about the stem of a
sycamore leaf which I had placed in the box
(as, when found, these baskets were suspended
by a thread several feet long, from a sycamore
branch near an arbor-vitæ). He worked at
this thread from five in the afternoon until ten
in the evening, making it as strong as possible,
as if to challenge a second disturbance. Then
he drew up the neck and kept quiet. Now
and then, the basket would shake, and swell
out to its fullest capacity. About nine in the
evening, I noticed the one which I had supposed dead moving. With sharp scissors I carefully cut off the very edge of the closed neck.
In about five minutes I saw him draw it gently
together. On the 7th, one of them pushed
his cast-off skin through the case, when I concluded that he was changed to a chrysalis.
But no, a little later the same day, he put out
a fresh head and shoulders from the bottom
of his sack, shook off the skin which had not
been quite freed before, and peered about him !
Then he retired, drew up the opening, as a
lady would her work-bag, and, as a caterpillar

I saw him no more. Fresh hemlock, pine, and
arbor-vitæ laid close to his basket seemed no
temptation to him to undraw those little
strings, and by closest watching I could not
see that he ate again after the change. The
basket would occasionally whirl violently, and
then remain perfectly still. Five segments
were the most it ever showed. The first
three rings back of the head are shelly in
appearance. In color it is a grayish-olive,
mottled with white, something like a tortoise
shell. The mouth and feet are an amber
brown. On August 13th, I looked within one
basket and found a very dark chrysalis. It
was quick in its motions, as was the caterpillar
and also the moth. On September 14th, two
of the chrysalides pushed out from the basket
and in less than half a minute with a little
bustling whirr the moths were out. They
were black with clear wings, which were
shorter than their long tapering bodies, giving
them a very curious appearance. The female
is wingless, as the Orgyia; white with an
amber-colored head, and would scarcely be
recognized as a moth. The antennæ of the
male are doubly feathered. Their basket
home is soft-lined and the neck both without
and within is free from sticks and soft as plush

to the touch.   The female moth never leaves her home.   This evergreen Basket-worm is doubtless the species Oiketicus, of Harris (p. 415) and which he says "is common in the vicinity of Philadelphia on the arbor-vitæ, larch, and hemlock."   I found them this year on evergreens at Ocean Grove.

XXI.

A BLACKBERRY LOOPER.

ON July 17, 1884, I secured from a black-berry a very curious "looper" cater-pillar. He was of a mulberry-brown color mottled and ringed, and his body shagreened. He had two pair of hind prop feet, and three pair of true feet in front. His head looked like a double hoof of a cow's foot. If he had been a gymnast or acrobat his fortune would have been assured. Any man who could stand with his feet against a tree, in a perfectly rigid horizontal position, an hour at a time, without moving, might well attract a crowd at a dollar an hour.

This gymnast exhibited free, and astonished you by the wonderful variety of his exploits, and stoical immovability from his position when taken. Now he was a stem to the black-berry. Again a handle to it. Then a syphon ; again an " eye " waiting for the corresponding hook.   Then a loop-and-link, as if he had begun to make a chain, and gave out on the second link.   Then he made a stiff bridge from one berry to another.

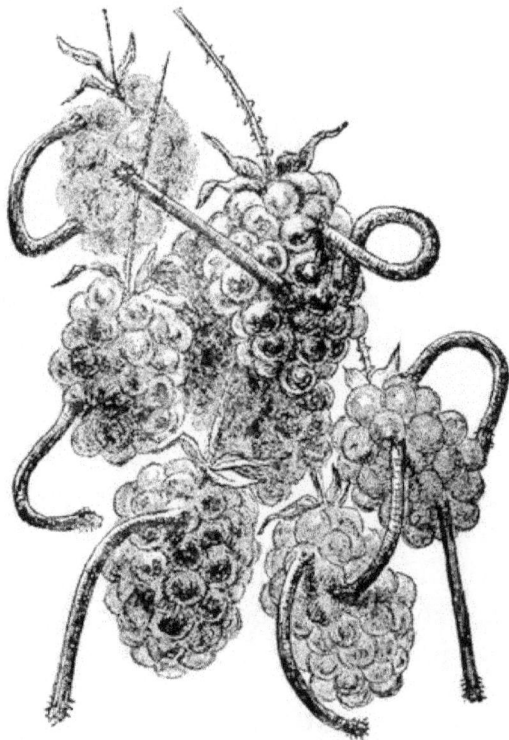

FIG. 63.   DIFFERENT POSITIONS OF THE BLACKBERRY LOOPER.

Each of his true feet is armed with an amber claw.   The mouth also is amber-colored and yellow.   It is difficult to see it, even with the microscope, as it appears as if drawn into its body, or neck, when not in use.

His little group of eyes, or ocelli, are plain to be seen, and he would peer forward in response to my watching, as much as to say :

" I have as much of an inquiring mind as you ! "

On July 26th the looper changed to a buff-colored chrysalis, very pointed at the end and having a dark-brown central line (interrupted) down the back.

On August 10th the chrysalis opened, and out came a most delicate pea-green moth, with white wavy bands on fore and hind wings, both of which were fringed. The body is a creamy, silvery white, the head and feet light amber. The legs are spined, one spine on each. The plumed, amber-colored antennæ are broad at the base and taper to a point. The eyes are *large* and of a sage-green color, with a dark circular ring, which appears like a pupil, near the centre, and which under a microscope gives you the feeling of being looked on with a responsive gaze.

The moth has been identified for me by Professor Lintner, as the *Nemoria chloroleucaria* (Gueneé), and is said to be distributed over the United States from Canada to Texas, and is no doubt far better known in its perfect state than as a blackberry-loving caterpillar.

XXII.

THE DRYOCAMPA IMPERIALIS.

HEARING a slight noise in my room one evening, I turned to look at the chestnut brown chrysalis I had long been watching, of the beautiful moth *Dryocampa imperialis*. Having lost the caterpillar of this moth the year before in making its change, I was very glad to see this fine chrysalis (which had afterward been sent me by a friend) at last show signs of opening. This was the first of May, 1880. As I looked a slight parting appeared exactly in the centre of the front of the head, giving a glimpse of the yellow color of the moth. The quickness of the parting and closing of this narrow thread-line, showing the rich golden yellow for an instant, was like the play of miniature " heat lightning." Watching it until after midnight, the chrysalis at last became perfectly motionless, and I left it, thinking it would not move again. To my

FIG. 64. THE DRYOCAMPA IMPERIALIS.

surprise, the next morning there was the same flashing of the little yellow line, which continued without any gain throughout that day, and half of the day following. Thinking it would not be able to break the thick shell, with a fine needle I carefully broke off some tiny bits from the side of the crack, and soon, with a mighty stir and bustle, the moth walked out.   But alas, never to shake out the rich purple and yellow wings ! Whether he would have finally succeeded in freeing himself from the chrysalis alone, or not, it is certain my assistance did him no service, and the beauty that " might have been," and which was partially revealed by the imperfect moth, only added to my second disappointment.   The next summer I received a fine specimen of the caterpillar, from a friend who had found it on its favorite button-wood.   I had scarcely time, after placing it upon a box of earth, to note carefully its sage-green color, reddish-tinged back, orange head and feet, white, green-bordered spiracles, and the six yellow, black-spined knobs on each of the wings except the first, before it worked itself rapidly out of sight, to make its change in the ground.

When the box was being opened the dinner-bell proved, for once, an unwelcome sound, but thinking (and wisely as the event proved) that *now* was the best time to secure him, I seized my pencil and made the following sketch before satisfying my appetite.

On my return from the dinner-table the surface of earth in my box was marked by a half-circular ridge, about the width of the caterpillar's body ; it had gone from sight to make its wonderful change. This it did successfully, and having slept itself into its spring suit (in which matter caterpillars have greatly the advantage of us), it came out of its prison in May, in its exquisite robe of yellow and purple, and with as much ease and celerity as if its ring-notched case was not to be thought of as an obstruction, when it was ready to give me its full-dress surprise.

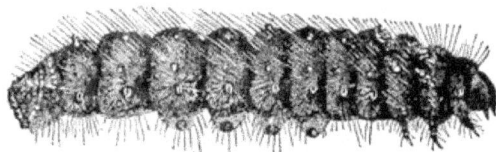

FIG. 65. CATERPILLAR OF DRYOCAMPA IMPERIALIS.

This large caterpillar feeds upon the Sycamore, and is found during August and September. Some of them are over three inches in length. They are of a peculiar shade of

green, in some cases with a faint reddish
flush, and occasionally a more rusty olive-
brown. The feet are orange-colored, the
spiracles double-bordered with white and
green. On the second and third rings are
two knob-like horns curved backwards, of a
bright yellow—the three pieces, shaped like a
triangle at the end of the body are also yellow-edged, sprinkled with small dots of orange-
colored knobs, and on each of the rings are
six thorned yellow knobs. There are a few
thin hairs scattered over the body, but so
sparsely as to be scarcely noticeable.

He is a gentle caterpillar, like the Polyphemus, but much more difficult to carry through
its changes successfully. One which was
given me by a friend, the past summer (August, 1889), failed to complete his change into
the chrysalis, although every care was taken
that he might do so. He had been brought a
long distance, and possibly received some in-
jury by the way. This caterpillar is rare, certainly in Pennsylvania—and about as difficult
to secure in an afternoon search, as is that of
the Royal Walnut. I have as yet never been
so fortunate as to find one, although sundry
protracted peerings into the leafy boughs of
the Sycamore on many a ramble may have

suggested to an on-looker the thought that he had encountered some one not very re- motely connected with Zaccheus. Probably the easiest way to obtain them will be found through the egg, by securing the moth itself. In the capture of a moth one should not lose this possibility of a bonanza by *"jarring" them at once*, in order to secure a "perfect speci- men." This thought came to me just in time to save my putting a fine female Luna moth into the cyanide jar as soon as caught, when, had I done so, I should never have been able to record my " Barrel Full of Lunas."

XXIII.

A BARREL FULL OF LUNAS.

ON June 22, 1883, a beautiful Luna moth was given me by a friend. It was the first living moth of this kind I had had, never having been so fortunate as to secure the caterpillar or its cocoon. Just as I was about putting it in the cyanide jar, the thought struck me that I might possibly secure eggs and raise moths of this beautiful species. Scarcely had I decided to keep it, before I noticed a cluster of eggs on the inside cover of the box in which it was brought to me. Here was a treasure indeed. And, in three days after, there were over thirty eggs in the box. They were dark brown, a little smaller than those of the Polyphemus, and biscuit-shaped like them, each having also a slight central depression. Most of them were lying in the form of a chain, in an almost regular connected line. On the second of July many of

FIG. 66. ATTACUS LUNA MOTH.

the eggs hatched, the young caterpillars being a light pea-green, a little less than an inch long. The spines were in clusters, like those of the *Io Saturnia*, those on the back having a purplish tinge. They began to eat fresh walnut leaves at once. In a day or two, little beaded knobs began to show, running lengthwise in rows. Some, which had moulted, had rich purple tufts on them,—four on the two front rings (two on each) and *one* on the last ring. The true feet were also purple, or purplish brown. The Luna caterpillars are easily kept. When the glass cover is removed they do not rush to get away, but eat on contentedly. They betray no snappishness like the tomato-worm. When the next change was made, the rows of crimson or garnet spots were much larger. Each crimson spot has a light-yellow border and a little tuft of hair from its centre. The true feet are dark-crimson—the false ones puffy and pea-green, like the body, and bordered at the clasping-edge with crimson. The head is green, marked on the front with crimson, and the mouth is crimson-tipped. When about to moult, the caterpillar fastened itself to the side of the glass by a netting of fine silken threads, head downward and bent forward, the true feet drawn to-

gether, exactly evenly, in pairs, giving it a
meek look, as if it were in the act of peti-
tioning for pity. It changed in about two
days. It was curious to watch these caterpillars eat, holding a leaf firmly with the three
pairs of true feet, and supporting itself by the
four pairs of prop feet, with their dull purple
or crimson sucker-like claspers clinging to the
stem. The leaf melts away before their rapid
cutting in a marvellous manner. The amber-
like spinnerets stand outside, and the jaws
work together sideways, the edge of the leaf
being guided by passing between two feelers
which hold it steadily in position as it disap-
pears beneath. When the worm was older,
the crimson buttons were shaded on the top to
light pink. The eight spiracles or breathing-
holes at the sides are shaded crimson (a puffed
line of yellow-green bordering them), running
lengthwise, and cut into lengths by each ring.
In the centre of each puff is a crimson dot.
When fully grown, the head is sea-green, as
also the V-shaped spot on the tail, which is
bordered with yellow, and ends with a brown
clasper foot, yellow-edged. The true feet are
black. The mouth is very elaborate. With a
microscope and a good stock of patience, the
exact number of these spots of crimson, which

so adorn the Luna caterpillar, were counted.
On the first ring, there are six ; on the second,
eight ; on the third, eight ; on the fourth, fifth,
sixth, seventh, eighth, ninth and tenth rings,
six each ; five on the eleventh, and four on
the last ring, making seventy-five in all. They
are larger than the same beautifully-colored
points on the Polyphemus caterpillar.

On the 21st of July, just as they were almost ready to spin up, I was ready to leave
on a vacation of at least a fortnight. Two or
three on that day had begun to make cocoons,
but several, a good deal smaller, must eat
some days longer. No one was to remain in
the house, and what was to be done ? To lose
twenty or thirty Luna moths was not to be
thought of. Had they been canaries, a friend
could have been asked to take them in charge.
But even the superior beauty of the crimson-
bedecked caterpillar might not bring it into
sufficient favor to secure the granting of such
a request. The problem was happily solved.
Ten of them were taken, in a wire box, on the
journey, and as walnut trees are not abundant
in the part of Massachusetts whither my way
tended, an extempore *silo* was made by press-
ing very closely a quantity of fresh leaves in a
tight tin box. This lasted the ten travellers,

and they each made a perfect cocoon against
the sides of the box. But for those left
behind? A clean barrel was secured. This
was papered inside and out with newspapers.
Then a large glass jar was filled with water,
and long sprays, freshly cut, of walnut were
placed in the jar, and this put in the barrel.
Then the caterpillars were at home, and by
covering the top of the barrel with a rather
fine wire sieve, they had plenty of air, and were
kept at home.

In about three weeks the well-formed
cocoons in the travelling wire box began to
open, much to my surprise, as I had supposed
they were to remain until spring. Reaching
home soon after, it was no small pleasure to
find not only the cocoons but several moths
already out. The contents of the barrel were
examined with no little interest, as well as a
glass shade which covered some which had
been placed under that with a bottle of water
filled with leaves. The moths in a few in-
stances had broken their wings, but many were
still perfect. They were perhaps a little
smaller, but not less handsome, for their
rather cramping experience. Very few had
died, and there were still some leaves left not
altogether shrivelled.

The Luna moth is of exquisite form, and delicate colors. It is a light pea-green, with edges bordered with yellow, and a brown edge to each fore wing. It is tailed, and has two handsome transparent centred eye-spots, of white, black, yellow, and a faint tint of red. Those of the hinder wings are round, and those on the fore wings are like an inverted comma. The body is white and covered with a soft, fine wool, the antennæ yellow and plumed, and the legs a purple brown. The colors in the eye-spots are so blended as scarcely to be separately distinguished without a glass, the whole appearing like shades of brownish pink.

The cocoon is made much as that of the Polyphemus, but is not attached so firmly to the stem or branch. The experiment of raising them is simple, and of special interest whether done at home or abroad.

FIG. 67. COCOON OF ATTACUS LUNA MOTH.

XXIV.

A LITTLE fluttering noise as I passed, last
February, a shelf where chrysalids are
kept under glass, revealed a spring, or, rather,
winter "opening." The first butterfly to appear
from among the many housed sleepers was
from a chrysalis long and carefully watched,
and which came out February 21, 1884. It
was the Cresphontes butterfly, and should be
a large and handsome one, but, alas! from
some unknown reason, he appeared with sadly
crumpled hinder wings.

If one has an unusually long chase for a
butterfly he has never had before, and breaks
his wings in taking him, it is disappointment
enough; but to wait, without even the excite-
ment of a chase, from November 15th to
February 21st following, and then have an
imperfect one, seems almost too bad. How-

ever, here he is, and wide awake, and, so far
as perfect, handsome. And " life " shall be
made for him as nearly " worth living " as possible. In fact, he looks as if he thought it
were, now, as he uncoils his long, black, three-
grooved tongue, and sucks the sweetened
water from the beautiful cups of creeping
evergreen *(pyxidanthera barbulata)*, which
seemed to come from the South to-day on
purpose to give this Southern butterfly a welcome. At any rate, not more than half an
hour after he left his cell, the postman left the
box of " moss " which came from Wilmington,
N. C., and its beauty and sweetness *must* atone
for his poor, folded-up wings. He sips eagerly,
and raises his front wings and sways his long,
over thirty-jointed or ringed antennæ to ex-
press his satisfaction. The Cresphontes is,
when of full size, with wings spread, from four
to five inches across. He resembles the Turnus butterfly in color and form, but the markings are different. I have seen but two—the
handsomer of these in the Lenox Academy
Museum, last summer, and one which flew into
a friend's house, on College Hill, Easton, Pa.
These caterpillars, for I had three of them
last fall, are very curious, and entirely different
in appearance, in that form of their life, from

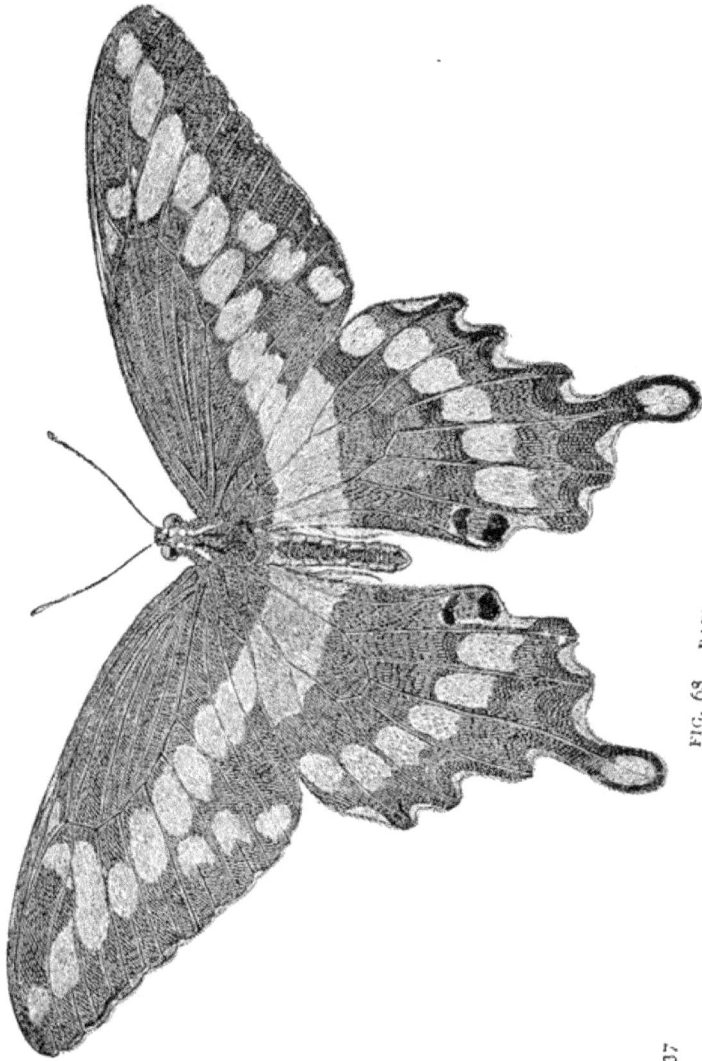

FIG. 69. PAPILIO CRESPHONTES.

the Turnus. They were found on a prickly-ash tree in the grounds near where the butter-fly above named was caught. They ate the leaves of the prickly-ash, but did not seem very fond of it. In reading what I could find of the Cresphontes, I learned that in Florida it lives upon the foliage of the orange tree, which is, I then noticed, classed in the same family with the prickly-ash. (See Gray's "School and Field Book of Botany," p. 81 Rue Family.)

So, thinking they might prefer orange to prickly-ash, I obtained sprays of leaves from an orange tree in a friend's conservatory, but they turned from it with contempt, as much as to say, " I know in what locality I am, and if I can't have my native air, I will not accept my native leaf." And here I must mention a fact noticed several times with much interest. It may not always hold, but has, I believe, in each case that I have watched. The caterpillar that is said to like several kinds of leaves, will *prefer* the kind on which it first found itself and began to feed upon. The *Saturnia Io*, found upon the corn blade, refused the dogwood leaf, which it is said to like ; and a Polyphemus, found on an oak, in Massachu-setts, turned away from the maple every time,

although those found on the maple ate that
greedily. No doubt, rather than starve, they
would take some of the other kinds which
they are said to eat, although I think the
Cresphontes would have starved sooner than
touch the orange leaves. The description of
the caterpillar I quote from my butterfly journal, written with the living specimens before
me.

" Oct. 15, 1883.—The shape of the Cresphontes caterpillar is very curious, and the
colors rich and velvety. It is hooded, the
hood covering much of the time its small,
olive-green head. The hood is ornamented
with round rings (of white or russet), four
round rings on the front edge and ten on the
lower edge. One of the three (smaller than
the others) has a moist, slimy look, and the
rings look more like little clear bubbles than
well-defined circles. There are six lavender-
colored, irregularly-regular spots on the back,
just above the white and bulging end. The
sides are grayish-green. With a microscope,
the rings show beautifully, and one wonders
at the amount of exact work in.so small a
space. The olive-green head has a white line,
which runs straight down the centre a little
way, and parts in a delta. It has a pair of

crimson horns, which do not show except when disturbed. The true feet are a clear, light olive-green, — the false feet grayish-green, fringed with white hairs, marked and mottled with small crescents.

" Oct. 16th.—One of the Cresphontes spun a few threads, and attached himself to the side of the glass box to change his coat. His head is small and black, and is meekly bent against the glass, not in sight, looking from above.

" Oct. 17th.—The Cresphontes keeps perfectly still against the glass. Watching him carefully with a microscope, I cannot see the least movement, more than if he were dead.

" Oct. 19th.—The Cresphontes still fixed against the glass. With the microscope I saw a most minute insect (not half so large as a period on this page), on his head, which annoyed him. Brushing it off with a feather, he threw out his crimson horns, and revealed well where they protruded. From a horizontal slit on the forehead (an almost imperceptible line), both issued from one opening, being joined at the base, in one short, crimson stem, which is close to the angular top of the head. I had doubted whether he could use these, having been so long suspended for his change ; but he did, readily."

At length the Cresphontes died, after leaving the glass, and soon both the others died. A friend then gave me a perfect chrysalis, formed at the same time, and which yielded the crumpled butterfly. The wings are jet-black above, with an irregular band of almost golden-yellow spots on the upper pair. The hind wings are bordered with yellow some distance from the scalloped and tailed edges.

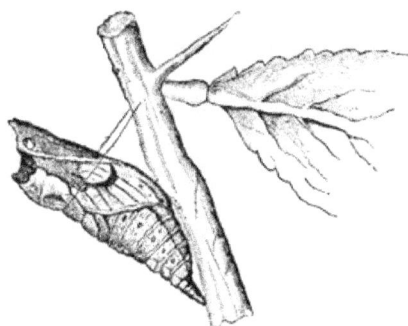

FIG. 69. CHRYSALIS OF CRESPHONTES CATERPILLAR.

The chrysalis is much like that of the Turnus in shape, and is suspended, like that, by a silken thread around the body.

Having sent the above sketch of my Cresphontes to a paper, it was noticed by a lady in Florida where this caterpillar is a well-known devourer of the orange foliage, and where there are often four broods during the year.

She very kindly sent me a box of the butterflies, so large and beautiful that I was well
repaid for the disappointment my poor de-
formed specimen had given me. The common name in Florida for this caterpillar is
" the orange dog," from a fancied resemblance
of its most curious head to that animal. When
in its native home the caterpillar is much

FIG. 70.   THE CRESPHONTES CATERPILLAR.

larger than the specimens I had obtained from
the prickly-ash, in Pennsylvania, where it was
evidently a new comer, and not to be found at
its best. In Florida it is found nearly three
inches in length. The gray and brown caterpillar, after feeding for a month, changes to
the chrysalis, and after a sleep of from one to

two weeks appears in the beautiful bright-winged Cresphontes butterfly. Its dull colors are said to resemble the bark of the orange tree so exactly as to make it difficult to be found, except upon close examination, a good example of the safety afforded to many insects by this conformity of color to their exposed places of living, while in a helpless state. When winged they can afford to triumph in the safety of flight, fearless of colors of a brilliant hue.

Since writing the above I have seen, by a report of an Ohio entomologist, the "prickly-ash" given as the "food plant of the Cresphontes" in that State.—[March, 1890.]

XXV.

A THOUSAND TO ONE.

TO every caterpillar its own secret. It can keep it well, but not forever,—truth will out at last. I can almost imagine one of them laughing at your surprise, as, after day by day you have carefully taken long walks to provide its special food, and watched it spin its patiently-wrought silken house, you look for the imago of the moth or butterfly you have " studied up," to appear, and, lo ! instead, a company of buzzing intruders—five, ten, twenty, a hundred ichneumon-flies *(Copidosoma truncatellum)*. No little suspicious-looking rice grains, even, carried around on its back (such as some caterpillars bear, to hint of disappointment beforehand), were to be seen on the back of the pale-green caterpillar secured from a stalk of wild-lettuce on the 24th of October. It was a fine-looking specimen of *Plusia brassiæ ;* and, as it was so late in the season getting ready for its change, special care had to be taken to select from among the already dying leaves of lettuce enough unwithered ones to satisfy it at its daily meals. However, as this was but for four days after its capture, it was done ; and then it mounted to the top of its glass prison, curled itself into the shape of a letter S, and began to spin threads of silvery-white silk back and forth around it, completing the covering while it was yet thin enough to disclose its zig-zag outline beneath the web. It was delicate pea-green in color, having two pale straw-colored stripes running down each side of a line of pea-green in the middle of the back, while on each side of this was a line of still brighter yellow, and each of the rings was so constricted as to occasion a corded appearance. The head was pea-green, like the body, with a small russet spot on each side. There were but two pairs of "false feet" (beside the prop-feet at the end) ; so, of course, it was rather an unusually strange "looper." Under a microscope the stripes appeared wavy, like watered silk ; and irregularly scattered over its body were tiny white dots, many of them bearing a short bristly hair, not to be seen except with the microscope. It was

10

about an inch and a half long, tapering toward the head; and this, with the curious
shapes it assumed when walking or feeding,
made it an interesting object to study. Sometimes it would lie straight along the stem,
but, if disturbed, would quickly loop itself,
and stay in its bent posture until reassured.

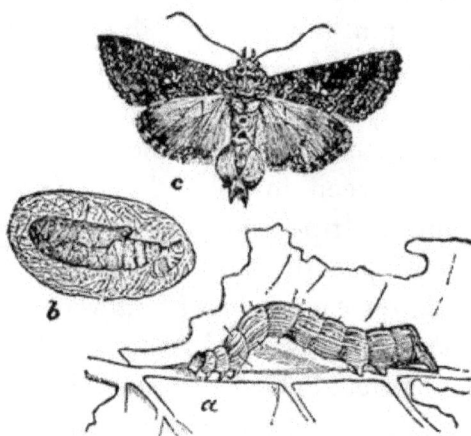

FIG. 71.   THE CABBAGE PLUSIA, PLUSIA BRASSICÆ: *a*, THE LARVA;
b, THE PUPA WITHIN THE COCOON; *c*, THE MALE MOTH.

We waited until November 18th for the
change, when we should see, instead of a
striped, halting, looper caterpillar, coiled up
in his silvery hammock, a beautiful tufted
moth. A curious change came, indeed, but
far more so than we had anticipated. At the
" opening " that November morning, no gray-

yellow-and-silver-winged creature appeared, as we surely had a right to expect, but instead, under the glass, fully one thousand brilliant tiny ichneumon flies. With black heads and iridescent wings (a shade of turquoise blue prevailing), this busy little cloud of intruders darkened and brightened the glass prison. In the first surprise of the moment the glass was lifted a very little, when dozens escaped around the edge. These were instantly brushed into a place of safety, and the rest secured by replacing the glass. The caterpillar's secret was out, and the task he had left for me was—counting. For who that has not seen it, is going to believe that from one caterpillar (after he has lived out his first stage of life, and built his resting-place for the next two) there should spring, as I have asserted, a thousand other lives ? So, after several days, when all the busy, darting gleaming rainbow specks were forever still, I took off the glass, put them on a white paper, and with the point of my penknife moved them off in groups of tens and hundreds, and, besides all that had at first escaped, there were by actual count eight hundred and thirty-two.

In looking up all I could find about this *Plusia brassicæ* moth (for I had seen only the

figure of it), I was interested especially in one
fact given in a number of the *American Entomologist* (1880), viz., that this caterpillar is a
veritable cannibal, and is quite ready, if its
legitimate meal of lettuce, cabbage, or turnip
be not at hand, to make a dinner off a neigh-
boring caterpillar of a different family, and
even to threaten the same unkindly office to
one of its own.   If the little ichneumon-fly
has happened to note this propensity, it has
surely had ample satisfaction in the way of
revenge.

The caterpillar of the Plusia is a great rob-
ber when found in abundance, as in many
places, eating cabbage, lettuce, tomato, turnip,
and especially celery.   It has been a great an-
noyance in Washington city.   So in Eastern
Pennsylvania we may congratulate ourselves
if they are so scarce as to prove, in a single
case. a treasure to the entomologist.

XXVI.

THE COMPLAINT OF THE CHRYSALIS.

THEY are in such a terrible hurry
 To see what I 'm going to be !
I 've heard them all talking it over
 But I fear that they never will see.

They took me from out my dark chamber,[1]
 Where the light strikes me now all the day ;
And if I don't move then they push me,
 To see if I 've died by the way !

As soon as my wings get some color
 And begin just a little to show,
Beneath my poor helpless brown cover
 What is hidden they 're crazy to know !

Dame Nature, my kindest of mothers,
 I *hope* she will see me safe through,
But I tell you she will not be hurried,
 Whatever impatience may do !

---

[1] The *cocoon* is often opened without harm to its enclosed chrysalis, that the changes of the latter may be noticed as it approaches the imago.

149

If you only would leave me in darkness,
  In quiet and silence to rest,
I 'd burst on you some pleasant morning
  In perfection of beauty full dressed.

But I think that last touch on my shoulder
  Has injured a delicate wing,
And I tremble to think of your waiting
  To welcome a poor blighted thing.

I should like just the chance once to show you
  How lovely a moth can appear
Who has slept undisturbed in his casket
  His little two-thirds of a year.

XXVII.

THE TUSSOCK MOTH.

I HAVE been trying to-day to feed a moth, or to find whether he has a tongue. Hearing a slight rustling noise coming from a shelf where the sleepers in several tufted felt-like cocoons had

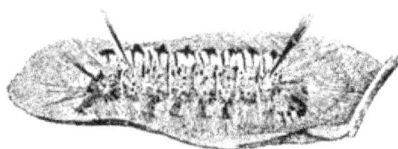

FIG. 72. HICKORY TUSSOCK CATERPILLAR. been taking their long winter naps, I looked to see if it were possible that any of them had been cheated by the unusually mild weather into the belief that Spring had come. Sure enough under two different glasses fluttered, this January day (Jan. 11,

FIG. 73. HICKORY TUSSOCK MOTH.

1880), two buff-and-white spotted Tussock moths, wide awake and ready for flowers,

while snow covered the bare branches of the
hickory trees, where their first life was spent.
Looking at the date on the paper with the
label, I find that the larger of the two moths,
(which are alike), was found, a full grown
caterpillar, in the previous autumn, September
13, 1879. Turning to this date in my Butterfly Diary, I find he was a yellowish-olive-
green caterpillar, with yellow brush-like tufts
on his back, a pencil of white hairs on each
side of the first ring, a pencil of dark hairs on
each side of the second ring, and two black
pencils from the last ring. All the feet,
"true" and "false," were of a clear amber
hue. The head was jetty-black, with small
white spots at the mouth. This caterpillar
draws down the white pencils of the first ring
so as to veil the whole head, which makes the
two dark or black pencils of the second ring
stand out like horns. The first ring has a few
small oblong yellow spots upon it. When
disturbed it instantly rolls into a round button-
like coil, remaining for some time perfectly
still. When all danger seems past, it as suddenly starts from its pretended sleep, and
walks rapidly as far as its prison will allow.

The caterpillar of the second moth was
found on September 16th, three days later

than the first. He was a pale lemon-yellow color, with an amber-colored head. Although I see no difference, upon the most careful examination, the caterpillars were thus slightly different in color. The four pencils in front, of the second caterpillar, were of a deep orange color. There were back of these two pairs of shorter white pencils, and the two from the last ring were also white. The tufts (like small square cushions) on the back, spring each from a black-dotted centre. The whole caterpillar has a soft and very neat appearance. The feet of this one were white instead of amber color. I have since raised many of these caterpillars and find that they vary in color, some being mouse-colored, some yellow, others gray, and others olive-green, and often those of one of these colors, on changing his coat will be found to go from gray to lemon-yellow, or from olive-green to drab, and yet the imagos or moths will all be of the same color, which is the exact shade of the hickory-nut meat, (a yellowish-brown), sprinkled with white dots. It is a quiet, gentle caterpillar after it once yields to its imprisonment, as a fixed fact. Until then it is unceasing in its efforts to find its freedom.

The cocoons of the Tussock moth caterpil-

lars are made entirely from their own hairs.
They are oval, as shown in the cut, and take,
of course, the color of the caterpillar in its last
stage.   Of the eight or ten now waiting in
their chrysalis state (December, 1889), some
are gray, some brown, and some of a delicate
purplish hue, but all will yield the hickory-nut-
meat moth.   Some of them are suspended
from the top of the glass, while others lie on
the paper at the bottom of the box.   They
differ in size, the largest being an inch in
length.   The opening at the end of the cocoon
where the moth makes its escape is so small as
not to be noticed at the first glance.   I said I
had been trying to find whether this moth had

FIG. 74.   COCOON OF THE
HICKORY TUSSOCK MOTH.

a tongue.   If he had it was
not to be tempted from its
covert by sweets, to which,
even when dropped upon
his mouth, he paid not the
slightest regard.   In fact,
he is proving that he
would rather die than eat, as he is now,
after several days' entire abstinence, very
nearly through his quiet little life.

Many of the Tussock moths described above
came out in January.   This year (March,
1890) not one cocoon has yet given up its

pretty moth, although the caterpillars of the
more than a dozen now waited for "spun up"
quite as early last fall as did those which appeared in January. Just twenty moths of the
Saturnia Io have come out in the same "chrysalis room" during the first half of this month,
and the last two weeks of February, as well as
two Polyphemus moths and two or three other
kinds, quite throwing the promptness of the
Tussock moths, this year, into the shade. But

> "They will not be hurried,
> Whatever impatience may do."

XXVIII.

WINGED AND WINGLESS.

"IN natural history nothing is small." This truth often strikes one when they direct the microscope to a little speck, hardly sure but that it may prove a grain of sand, to find a perfect insect, beautiful in form and adorning.

There is no caterpillar (perhaps with one exception) more handsome, to me, than a quite

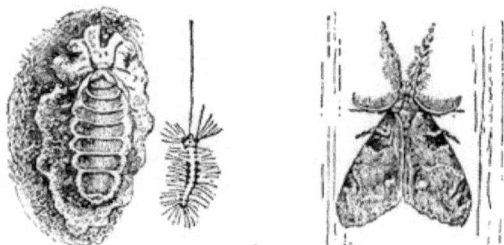

FIG. 75. FEMALE (WINGLESS) OF   FIG. 76. ORGYIA LEUCOSTIGMA
ORGYIA LEUCOSTIGMA.   MOTH.

common one found on the maple, or willow, and often on the rose leaves, in the early summer. (And here I may note that, lest I should not *fully* appreciate its beauty, and also that I

might understand the value of a worthy setting, one of these caterpillars walked slowly
over the satin crown of a lady's hat, immediately in front of me in church one Sabbath
morning, a perfect picture of beauty, while the
haste with which it was brushed from its well-
selected promenade ground by one who saw in it
"only a caterpillar," proved that in some eyes the
" setting " may be of more value than the gem.)

This caterpillar is not a very small one after
all, being an inch or a little more in length ; but
a microscope is needed
to reveal fully its many
special points of beauty.
Of the several which I
have tried to watch
through their caterpillar
life into the perfect
imago, I have but one
left, and for fear he too
will die, with his caterpillar frock on, I will
give his portrait with my pencil, although you
may have brushed him from you hastily, after
many a summer ramble, unstudied and igno-
rant of his beauty. He is one inch in length,
and his prevailing color is a delicate but bright
lemon-yellow. This forms a fine groundwork
for touches of peculiar beauty.

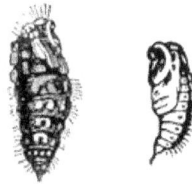

FIG. 77. · CHRYSALIS AND FE-
MALE MOTH OF ORGYIA
LEUCOSTIGMA.

His head is a pale coral-red.   His small
mouth jet-black.   The top of his head and his
sides are covered and fringed with a few soft
straw-colored hairs ; those of the head, bending forward, giving the coral the appearance
of being shaded with yellow.   If he were not
so restless it would be more easy to give an
exact description of him.   Moving his tufts
of hair backward and forward, it is about as
easy to count his twelve rings as to count a
long train of cars, in good motion.

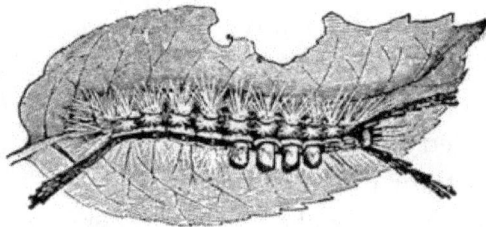

FIG. 78.   CATERPILLAR OF THE ORGYIA LEUCOSTIGMA.

From the first ring there springs a long
wavy pencil-plume, just back of the coral head ;
a brush of black hairs, shingled in sets of two
or three different lengths, and each of these
hairs feathered like an arrow at the tip.   From
the fourth, fifth, sixth, and seventh rings, there
rises a very full, thick, even brush of soft yellow hairs, like four miniature oblong clothes-

brushes, laid one after another along his back. On the top of the eighth ring is a flat spot of a beautiful crimson color, while the ninth and tenth rings have each a little crimson ball upon their top. From the centre of the eleventh ring, rises another beautiful pencilled plume, and the twelfth is finished with a delicate fringe of fine brown hairs.

Like other caterpillars he changes his coat three or four times. The long plumes coming off with the discarded coat, while new, fresh, and longer ones are ready at once to take their place. After wondering, while watching this change, how long it would take for these shingled plumes to grow, after the old ones were thrown off, what was my surprise to see them slowly rise up, fully formed and handsomer than those laid aside a moment before !

He is a very restless caterpillar, and probably his dislike to imprisonment is the reason he so often fails to reach the chrysalis state. The one whose picture was taken above, died, but another, a fine specimen, was soon obtained, and placed in a larger glass box—one nearly a foot square. Here, with plenty of food—rose leaves and horse-chestnut leaves, (of which they are very fond also), he seemed content, and grew finely. Being ready to leave home for a fortnight's absence, I placed a large bottle of water in the box and put a branch of maple in it, judging this would afford him tolerably fresh food as long as he would wish to eat. Returning from my visit, and going to see how the prisoner was getting on, lo! he was gone. The maple leaves, however, had been pretty thoroughly devoured, and thinking he might possible have pressed his slender body between one of the ribbon-bound and rather lightly fastened sides of the box, I searched thoroughly for the little truant. The box stood upon a mantel, above which hung a heavily framed portrait. This was taken down and lo, on the back, in one corner of the frame, was a small, very thin, and odd-looking chrysalis. It was gray in color, and formed of hair, with little rough spots upon the sides and top. The plumy pencils had gone to form the cocoon! Not many days after, there was a small opening in one end of the cocoon, and on its top was a clear white glassy looking, or frothy appearing, substance which looked something like a few crushed glass beads dropped in glittering pinches upon the gray cocoon. Presiding over this mass, (of what was really a cluster of eggs, covered with a frothy substance), was the queerest little apol-

ogy for a moth ! "Can this be all, after all
my watching ? " I said. I thought it an unfinished bit of Nature's work—a deformity, and
had well-nigh brushed the whole away in my
haste, when lo ! a pretty little stranger moth
flew by me, hovering near the chrysalis. *This*,
I thought, might after all, be the true moth
from my handsome caterpillar, and I determined to prove it. Confining him, I kept
watch for another caterpillar, and was fortunate enough to secure two or three large specimens ; and, in their transformation to find,
first, another of the queer, almost wingless,
whitish moths, *and*, also, two, the exact mates
of the pretty gray-winged one I had before
caught and still held a prisoner. Winged,
and wingless ! The upper pair of wings to this
male moth were banded with wavy lines of a
darker, ashen-gray, and had a small black spot
near the tip of each wing, with a very small crescent of white near the outer edge. The back
of the moth was tufted handsomely, but all of
ashen-gray. No hint of scarlet, coral, or
crimson. The glassy frosted eggs again appeared by the wingless moth (who never left
them), on the cocoon's top, and opened at
length to release minute specimens of my coral-
headed, pencilled-plumed caterpillars. These,

11

not being noticed soon enough to receive their rose-leaf meals, soon lived out their little lives, but not until I had taken the caterpillar of the *Orgyia Leucostigma* safely through its little round of insect life. Certainly the beauty, in his case, lies in the first stage of his existence ; although the gray-banded crescent— marked Orgyia has a quiet beauty not to be overlooked.

XXIX.

A RACE FOR LIFE.

NOT between man and man and not between animals, but between a plant and
an insect. " The gooseberries look splendid-
ly this year. I
do not believe
that they will
be attacked by
the persistent
little enemies
that r u i n e d
them last year."

A day or two
after this cheer-
ing prophecy
last spring an-
other examina-
tion of the flour-

FIG. 79.

ishing bushes revealed the unmistakable enemy
in full force. On May 16th, three of them

163

were brought in and put under glass. The next day two threw off their coats and the third soon after, although eating up to the last moment, as is their greedy way. As every thing was early last year, so these false caterpillars were some days earlier than the year before. On May 20th of that year, they were fast putting on their last coats. Some were put under glass on that day and carefully watched. They had jet-black heads. The first ring back of the head was yellow, and there was a yellowish ring near the tail. The rest was a bluish-green, and the whole spotted thickly with black dots looking like little irregular drops of black sealing-wax. The feet were black. In crawling over the leaves one came near impaling himself on a thorn. He held back his head, sphinx-like, and considered the matter carefully, concluding to take the leaf and give the thorn a wide berth.

The next day (May 21st) one of these changed his coat and came out in a pretty and more spring-like suit of soft pea-green. He has the advantage of the leopard, for not a spot is to be seen on his new attire. The yellow bands are there as before, but not a vestige of black. The head is yellow, and the next two rings a brighter yellow, and also the third from the last and the last rings. The feet are also a very light clear green, almost colorless.

Keeping a close watch on the second it was easy to see him change his coat, which he accomplished in about five minutes. The head was first freed, and the old coat slipped back, aided by constant movements of the head and the fore part of the body already freed. He was ready to eat a fresh meal almost as soon as he was released.

After he changed his coat he spun a small yellow silk cocoon, almost transparent, drawing over a notched lobe of the leaf, half hiding it from sight. Whether it would come out a moth or a butterfly I was uncertain, having then never seen a description of this caterpillar. He did neither. On the third of June he came out a fly, with four transparent, netted wings, black head, and yellow body, with seven-jointed antennæ. There are some black spots just back of the head. There is bronze-like gloss

FIG. 80. CURRANT SAW-FLY.

to the clear, pretty wings, the legs are bright yellow, and the tips of the toes black. A careless observer would not think of his being more than an ordinary house-fly dressed up a little for an afternoon call.   Surprised at the shortness of the life of this caterpillar, I thought it strange they should be able to destroy the gooseberry leaves so completely.

FIG. 81.   CURRANT LEAF EATEN IN CIRCULAR HOLES
BY THE SAW-FLY.

The caterpillars soon seemed to be all gone; again the gooseberries threw out fresh young leaves and seemed determined to get the upper hand, but their triumph was short.   Very soon

the new leaves were bordered with the unmistakable black-spotted rim, a second brood left
the bushes bare, and not being satisfied with
their full meal adjourned, by way of dessert, to
the current bushes where, after a short stay,
their little cast-off dotted coats could be seen
all over the twigs and their yellow heads busy
making small crescents in the currant-leaves.
Not long after, their cocoons were spun and
they were snugly stowed away to await the new
leaves of another spring.

There are some kinds of caterpillars, and
these are among the number, which birds
avoid, and so if any one is to come to the
rescue of the gooseberries it must not be left
to them. But after watching them through all
the windings and changes of their curious little lives, and forgiving them for robbing me
two years in succession of gooseberry tarts and
currant pie, I will leave this part of the matter
to certain books where the secret of their ex-
termination may be found. According to one
of these books, they have been in this country
since 1860, when they were imported from
Europe into nurseries in Rochester, New York,
and are known by the name of " The Imported
Currant Saw Fly."

XXX.

THE BULRUSH CATERPILLAR.

AMONG the most curious productions of New Zealand is the singular plant (called by the natives *Awheto*), the *Sphæria Robertsia*, or bulrush caterpillar. If Nature ever takes revenges, one might imagine this to be a case of retaliation. Caterpillars live upon plants, devouring not only leaves, but bark, fruit, pith, root, and seeds; in short, every form of vegetable life is drawn upon by these voracious robbers. And here come a little seed that seems to say: "Turn about is fair play," and lodges on the wrinkled neck of the caterpillar, just at the time when he, satisfied with his thefts in the vegetable kingdom, goes out of sight, to change into a chrysalis and sleep his way into a new dress and a new life. A vain hope. The seed has the situation. It sends forth its tiny green stem, draws its life from the caterpillar, and not only sends up its little shoot

with the bulrush-stem capped with a tiny cat-
tail, but fills with its root the
entire body of its victim, changing it into a white pith-like vegetable substance. This, however,
preserves the exact shape of the
caterpillar. It is nut-like in substance, and is eaten by the natives with great relish.

A friend who has recently
spent some months in New Zealand brought me the specimen, a
drawing of which is here shown.

There are other cases of this
vegetable retaliation, but none
so curious as this of the bulrush
caterpillar. The larva of the
May beetle is attacked by a fungus which grows out of the sides
of its head ; but while this growth
destroys the life of the larva,
it does not change the larva
into a vegetable substance.

A near relation of the mur-
dered caterpillar is the larva of
the New Zealand swift moth,
upon whose tapering head sometimes appears a similar growth,

FIG. 82.

which feeds upon the life-blood of the caterpillar until it dies from exhaustion.

A very curious sight must be one of these
heavily-burdened crawlers moving along with
the banner that announces its doom solemnly
floating above it.   For, when the young cater-

FIG. 83.   LARVÆ OF THE NEW ZEALAND SWIFT MOTH.

pillar bears this growth upon its head, it heralds
the slow but certain death of the overloaded
insect.

A transformation as curious, perhaps, in an
opposite direction, is that of the insect *Drilus*,

which, in its larva state, lives upon the snail
—animal life drawn from animal, instead of
vegetable, substance. This beetle larva, with
its sucker-like feet, attaches itself to the shell
of the snail, watches its opportunity, and slips
inside. It lives upon the snail (sometimes
using three snails before changing to the
chrysalis state), and then, after it has finished
its last meal, it closes the door of the last shell,
and sleeps into its winged life. If insects
think us cruel in putting out their little lives
rather roughly, or if they complain that sometimes revengeful seeds change them into
miniature " caterpillars of salt," as it were,

Just let them study how they treat each other,
And learn more tenderness each for his brother ;
How innocent the small ant-lion,—sleeping
Beneath his pit of sand, while slowly creeping

Upon its edge a little ant comes near him,—
Then quickly, ere the ant has time to fear him,
Seizes his prey (the small deceitful sinner !)
With no compunction, for his stolen dinner !

The dragon-fly, in gauzy lace, and airy,
Sailing about like some delightful fairy,
Cares he what beauties butterflies embellish ?
He darts upon, and eats them with a relish !

In spite of all, if cruel still they style us,
Just let them think upon the thieving *Drilus*,
Who helix-back is very fond of riding.
And also into neighbors' homes of gliding.

And takes his meals without thanks to the donor,
Sleeps in his house and lives upon its owner.
Three rides he takes, three little homes up-breaking ;
Of three poor snails three travelling-pantries making.

A fortnight lives in each, the third one keeping
Quite to himself, at last ; and soundly sleeping,
Waits for his change—new life in some fair garden ;
But quite too late to ask the poor snail's pardon !

The singular change in this curious caterpillar is thus described by the friend who
brought me the above specimen, Rev. J. W.
Walker of Liverpool, and presented by Mr.
S. J. Capper, President of the Lancashire
and Cheshire Entomological Society at a
meeting of their society :

"This singular arrangement comes to pass
in the following fashion. When the caterpillar buries itself in the ground to pass into
the chrysalis stage, the minute spores of the
fungus find lodgment in the neck plates of
the caterpillar. There they vegetate, and
strike root inside the horny case of the animal,
living on its tissues. The animal dies, forming simply a root for this plant, which thus

lives on flesh. The bulrush attains a length
of about ten inches, its apex, in a state of
fructification, resembling the common club-
headed bulrush of our own ditches. When
fresh, these plants taste like a nut, and are
eaten by the natives, who also burn them and
use them for tattooing. When newly dug up,
the caterpillar's body is soft, and on being
divided longitudinally the intestinal channel is
plainly seen. The vegetating process com-
mences during the life of the caterpillar, for
decomposition has not set in, nor is the skin
expanded or contracted in any way. This
forms one of the most extraordinary freaks of
nature in the connection between animal and
vegetable, and is perhaps unequalled in the
annals of biology."

XXXI.

A BEADED CATERPILLAR.

A VERY pretty, small, smooth caterpillar was given me last fall, September, '88, found on some flowers of a bouquet, so that its special food could not be identified. A few days later, I was fortunate enough to find one like it upon a spray of golden-rod. These caterpillars were of a *seal-brown* color—all one shade of brown, very velvety in texture, about an inch and a quarter long, and cylindrical ; of one size throughout and about one third as large as an ordinary pipe-stem. What was my surprise on looking at the one given me (which I had placed under a glass), half an hour later, to see a very different looking caterpillar. Down each side was a row of white beads, perfectly symmetrical, and of a pure milky or chalky whiteness, globular and about the size of an ordinary pin's head.

Taking a magnifying glass I watched it with

interest. Presently one of the white beads disappeared, then another and another twinkled out, until lo ! the plain seal-brown caterpillar again. Touching it with a little stick, out came a bead here and there, their irregularity giving it a most curious look. I tried the second caterpillar, and it also threw out the chalk beads. I saw they were from the spiracles, but had never seen such a phenomenon before. On writing to Prof. Lintner about them, he replied that he had not observed any thing of the kind. The caterpillars, doubtless, may be found upon the golden-rod, as they ate of this plant and no other, which was tried, but neither of them (probably from having been disturbed too much in order to watch the coming and going of the beads), made a chrysalis. They are now small dried specimens, but remains of the white dots are visible upon them.

A few days since, I was interested in coming across a hint of this kind, found in an old encyclopædia (Rees, Art. " Stigmata "), in the following sentence. Speaking of experiments by Malpighi upon stigmata, he says:

" Mr. Reamer repeated his experiments, and concluded that these apertures served only for the *in*spiration of the air, which the caterpillar

afterwards *ex*pired through the whole super-
ficies of its body, because he could never
observe that any bubbles of air were ever
driven out of these stigmata ; but Mr. Bonnet,
on the contrary, *having seen bubbles of air*
coming out of these openings was led to infer
that the *in*spired air was also *re*spired or dis-
charged through these same orifices."

Now, if in the case of these two caterpillars,
it had been *merely* " bubbles of air," would not
the beads have been *glass-like* in clearness, instead of chalk-white ; and could they have
remained perfectly globular so long, as some
of them did for several minutes ?   These
questions it would be interesting to have
answered, as well as to learn the complete
history, in the imago, should any be able to
secure a specimen from next autumn's golden-
rod.

XXXII.

ATTACUS CYNTHIA.

O N March 21, 1879, I received from a
friend in New York a box containing
nine cocoons. When I first looked at them I
thought they were cocoons of the Attacus
Prometheus, so much, in all but their size, did
they resemble those familiar cocoons, many of
which I had watched open in the years before.
Like those of the Prometheus, the cocoon is
made with the leaf on which it feeds drawn
partially or entirely about it, and this is
securely fastened to the tree stem or branchlet;
but often upon so small a stem that this tight
winding does little toward the safety of the
chrysalis. Many fall from the tree and are
blown about, until, as one writer says, "the
streets of the cities in which they have become
wild are often strewn with such cocoons, which
get trodden on and destroyed." On cutting
open the cocoon of a few of my nine I found

FIG. 84. ATTACUS CYNTHIA: *a*, EGGS; *b*, LARVA; *c*, COCOON; *d*, CHRYSALIS; *e*, FEMALE MOTH (after Riley).

the chrysalids of a dark yellowish-brown, and in shape much like those of the Prometheus.

The first of these cocoons gave up its beautiful imago on the 24th of April, giving me a most pleasant surprise in a large moth of great delicacy and beauty of coloring. The body was thick and looked as if made of soft dark cotton with a close dotting of white tufts over it. The under side of the body was white tufted, on a yellowish-colored ground. The wings are of a yellowish-green hue, with variable markings, among which lavender is prominent. There is a crescent upon each wing, and a line of white edged with rose-color running across them gives it a striking and peculiar beauty. The moth varies in the time of waking from its chrysalis sleep, some remaining in their flossy-lined cocoons a much shorter time than others. The first one that came out of those sent me was a little over a month in making its exit.

The next moth of the " nine " came out on June 8th. It was handsomer in shading and colors than the first. The prevailing color was a beautiful shade of olive-green. The four crescents were a very light lavender color with a lower border of white and a yellowish olive-green. Through the centre of each wing was a wavy band of darker olive-green. But one must *see* to appreciate a moth with so many colors and such varied markings.

One more of the "nine," Cynthia's, came out a perfect moth, but, unfortunately, in attempting to escape in securing his liberty he rushed to his death. How the glass lid of his prison got moved enough for his escape I never knew, but I found him caught and marred (injured so that he died) in the partially-closed window of the room. How he was caught was also a mystery. The only solution that could occur to me being that he alighted on the frame of the sash, and was not noticed by the person who opened the window too suddenly for his attempted escape.

Through care of many other moths, and absence from home, the record ends here. One of these moths was secured from a cocoon given me, on June 20th of the next year, since which time I have not been fortunate enough to secure any.

The Cynthia is a native of Japan and China. It was introduced into France over thirty years ago, and many attempts have been made to use the floss of the cocoon in making silk, which have proved partially successful.

XXXIII.

ALTHOUGH the large gayly-colored butterfly, *Papilio turnus*—(as the best American butterfly-knower, Mr. W. H. Edwards, of Coalburgh, W. Va., says), " inhabits all sections of the United States, from the Atlantic to the Rocky Mountains, and from Maine to Florida and Texas," it had never

FIG. 85.   CATERPILLAR OF PAPILIO TURNUS.

been my good fortune to meet with one until I saw these flying over the large tulip trees of Eastern Pennsylvania. Their brilliant coloring attracted my attention, but their flight seemed always so high that after many attempts, I gave up the hope of securing one in the ordinary way. Had I known or thought of the caterpillar, or of the tiny egg, with its little silken hammock ready almost as soon as hatched, on

the upper side of a tulip leaf, I could sooner
have had possession of what I so much desired.
Or had I known that, when attracted by its
favorite flowers, the "blossoms of the wild
plum," (for which I give it great credit as
choosing almost the sweetest flower that
blooms), or hovering over beds of phlox, or
patches of red clover, it was so lost amid the
sweets of its eager meal that it could "be cap-
tured with the utmost ease," or that at any
given time or place it could possibly have been
found so abundant that Mr. Scudder (a very
careful and truthful scientist) could assert that
"sixty-nine of these butterflies had been caught
between the hands at one grasp!" I certainly
should not have paid ten cents a piece for two
or three broken-winged specimens brought me
by a little boy hired to secure the prize for
my collection.

However, a broken-winged butterfly, is better
than none, and studying from these, their marvellous beauty made me but too glad to learn
that the caterpillar lived chiefly upon the tulip-
tree leaves, although it did not despise many
other varieties of food. Fortunately there were
three of these handsome trees at our own door,
but I had never seen the butterfly about them.
I had only seen it, in long rambles, darting

FIG. 86. PAPILIO TURNUS.

183

through the upper boughs, (and often above all the boughs) of the tallest tulip-trees on the banks of the Delaware. Now for a search at home. Looking carefully among the lower leaves of the tulip, no egg was to be seen; (it needs a trained eye to see a small egg on a high bough !) but what is this singular looking caterpillar, walking toward me on the pavement near these trees ? I capture him and say : " Perhaps he may show me yet my *Papilio turnus.*" It was true prophecy. He was a dull looking and curiously marked little fellow, with a figure " 10 " always plainly marked on about the third ring on each side of his dark brown body, not far from the very large head, and giving him a quaint appearance, as suggestive of an inquiring eye.

My first Turnus caterpillar made his chrysalis September 20, 1878. He suspended himself much as the Asterias does, with a slender loop supporting him against the side of the box. The chrysalis, also resembles that of the Asterias; is a yellowish-brown in color and rough in texture, with his head prolonged in two ear-like points, and a similar projection a little below the head in front. [I have four of these now waiting (March 20, '90) their exit, after a six months' sleep.]

This first chrysalis of '78 opened on March
12, 1879. A large and handsome butterfly,
with no rude marks of boy capture marring his
perfect sunny wings of bright black and yellow.
Since then I have had many such openings,
but none ever gave me a more welcome or
highly prized Turnus than this. I found a fine
caterpillar of this kind in Sing Sing, N. Y., in
the summer of '82, and watched it make its
chrysalis on the 21st the following September.
Another which I studied more carefully a few
days later, as he made this change, gave me a
two hours' interesting study with the microscope on the evening of the 23d of September.
More than once I thought he had died from
his perfect stillness after efforts to effect his
change. On each side of this caterpillar, I
noticed (without the microscope, but very
plainly *with* it), three spots of vermilion sur-
rounded by a cluster of brown dots. Another,
not yet ready to change, although suspended
against the side of the box, had a band of
orange-yellow around the third ring. The one
with vermilion spots I saw finally throw off
his coat as if glad to be rid of it. The first of
these chrysalids opened on April 6, 1883.
Yellow and black dashed with spots of bright
orange. On April 16th, another opened, large

and handsome as any yet secured.   From two
or three of the Turnis chrysalids came a curious
Ichneumon fly.

It is a singular fact that the female of this
butterfly in climates warmer than that of Pennsylvania are often almost entirely black. These
are described in Mr. Wm. H. Edwards' full
account of this butterfly in his beautiful work
on "The Butterflies of North America."   He
speaks of this change of color as "without a
strict parallel among butterflies."   Another
butterfly-collector says that "in Georgia half
the females of the Turnus are black."

In all that I have seen, the male and female
Turnus butterflies are yellow and black and
very much alike in their appearance.

Why there should be this curious change of
color in those of the Southern climates is not
satisfactorily accounted for; and although several ingenious "suppositions" have been given
it is still left, by the best entomologists among
the mysteries which cannot be explained.

XXXIV.

(For illustration, see preface.)

IT is said by Harris, in his "Entomological Correspondence," of that most singular genus of moths, the Limacodes, that "they remain a long time in their cocoons, or in earth, before turning to pupæ." To this fact the student of entomology will give a ready assent !

In his larger work, Harris says of this *Limodes scapha:* "My specimens generally died after they had made their cocoon, and, consequently, the moth is unknown to me." Why he should use the word "generally" is a question, when if *one* only had not died, the moth might have been known to him.

By substituting *always* for "generally," my experience with the *Limodes scapha* is told.

Yet there is too much of curious interest in his first and second stage to omit a sketch of what, so far, must be but two thirds of a little life. He has certainly given me enough trouble, by way of watching and waiting, to warrant me in taking so much of his life as I can, especially as it is in a way which he cannot *feel.* He is, as a caterpillar, correctly figured (from a photograph which I had taken of one fine specimen) in the preface, where he is compared to the old-fashioned " beech-nut box." Harris saw rather a likeness to a little boat, and so "named him *scapha* (a skiff)."

As the change to a pupa *usually* takes place within a few days after the caterpillar has made its cocoon, one, before learning that the Scapha often remains for months in his little round parchment-like home, before throwing off his caterpillar coat, might easily give him up as dead, and throw away what would have paid for a little longer waiting. This I have done with one or two other kinds of caterpillars, who have the same habit, learning afterward my mistake. Some caterpillars lie in their cocoons through an entire winter, and then change into a chrysalis and finally come . out into their perfect state.

In a walk down a shaded lane, in September, 1879, I found my first *Limacodes scapha.* It seemed at first uncertain whether it was a raised place in the leaf, from some insect's sting, or something that was alive. I took it home, and, even with a microscope, could hardly determine whether it was animal or vegetable. At last it moved a *little,* not its position, but only a little contraction of the body, keeping almost as still as the leaf throughout the day. After the gas was lighted in the evening, I sat down with my microscope for a good look, and lo! the Scapha was just changing his coat. He was very little altered in appearance. He was a delicate pea-green, with a light spot of grayish brown on the top of his back, which slopes up to the centre and down again, like a water-shed. Just before he changed his coat, he puffed up and swelled like a puff-ball. He seems to be stomach-footed, like the saddle worm. He is marked by horizontal cross lines, and there are two triangular spots between each of the two lines before the middle line of his body, and two after it. He puffs out now and then, until he resembles a globe-fish. There are irregular brownish-gray spots on the sides, below a light central line, running through the centre from the broader front nearly to the apex. Back of the centre on the top, but on the backward slope, are two smaller grayish-brown spots. From 9 A.M. to 8 in the evening he has not changed his position on the sumach leaf, on which he was found. Harris gives the "walnut" as the food of the Scapha. This Scapha began to eat on the 23d (the day after he changed his coat) from an apple leaf. On October 5th the Scapha was very restless. He was now a pale straw color, or more nearly cream color. He walked about softly, looking as if gliding on water, like a little fairy-boat. The two spots on his back, just below the centre, look like two miniature lakes, and the two near the end are pear-shaped. The sides appear as if crimped. There are two small raised points on each side, at the centre line, very curious in appearance, which have showed from the first as a noticeable feature, and suggestive of eyes. I had thought they might be, when a friend, to whom I showed it, asked at once : "Are those little points eyes?" It would be no queerer place for them than the snail has for *his*. His head, for the most part, is entirely out of sight. When he puts it out (as a turtle does from its shell), it is round, of a light ochre-yellow brown, and has two pairs of feelers, one pair a little shorter than the other.

The Scapha has eight distinct breathing-holes, or spiracles, nearly round, and very dark brown or black, and about the size of a small period in fine print. This description, from my butterfly journal, ends with the true remark : " It seems as if, with a microscope, I might write an hour longer and not tell all his markings."

On October 9th he began to spin a slender cocoon. After beginning a very gossamer-like hammock, he stopped work and remained three days quite still, and eating nothing. More than once I was "sure" he was dead. Then he would move again, slowly rocking his tiny boat from stem to stern. Then he would draw in his head, and seem to be making an effort to change his coat, once going clear over in a funny summersault in the attempt. He was now orange-colored, and shrunken into hills and valleys. Towards evening he crept around to the little floss-silk tent or hammock, which he began and discarded some days before, and attempted to join a few floating threads for his cocoon, although his internal resources must have been limited, as he had not tasted food for four days! A magnificent example of patience and perseverance, far exceeding that of Bruce's spider, which had the physical strength *with* the patience, while the poor Scapha had to supply both. A good lesson from the little Limacodes, even if the *first* period of his triple life should be all he will ever attain. October 10th and 11th he was still at work, walking up and down and all around his box, working away toward his change. He had earned the right to call this "a changing world"; the only fear was that he would not find it big enough for him to change in!

October 13th.—Such an exhibition of *life* as the Scapha gives is seldom seen. Moving about, yet not having touched a leaf for a week and a half! The last record of him, *alive*, was on October 17th. Then he gave up his tent-working, and soon his struggle for life. Since then I have had several of these Limacodes, and some of them have made a nice plump cocoon, but never yet has one opened. On breaking the cocoon of one, after there was no hope of life about it, I found the shell very similar to that of the egg of a bird. It is nearly round, a dark brown in color, and smooth itself, although usually having a loose

dark flossy substance around it. I have now
(March, 1890) three or four well made Scapha
cocoons, and the wish to see them give up the
imago is in due proportion to the interest with
which its caterpillar life has been watched.
Although Harris had failed
to see the imago, and many
others have had the same
experience, it has been safely
brought through its changes,
and Packard has given the FIG. 87. SCAPHA MOTH.
figure of the perfect insect, from which the moth
here is taken, in his "Guide to the Study of Insects," p. 290. He says of the moth: "It is
light cinnamon brown, with a dark tan-colored
triangular spot, lined externally with *silver*,
which is continued along the costa" (or outer
edge of the wing) "to the base of the wing,
and terminates sharply on the apex." A living proof, with its silver finish, that *riches* will
still "take to themselves wings, and fly away."

A little after finishing the above sketch
(March 28, 1890), to my very pleasant surprise, I found under a glass in the chrysalis
room *my first* Limacodes moth! He had
stolen a march upon me at last, and, without
observation, had shaken out his wings of

"silver and cinnamon brown," lived his brief life (without the welcome and attention that would have been gladly given), and fallen asleep on the floor of his little prison; his empty brown cocoon and its little round lid lying beside him. He is, however, a prize in himself, and a herald of hope for the four still unopened brown balls, for which he has secured a closer watch for their possible opening.

XXXV.

O F all the *curious* caterpillars it has been my
good fortune to see, the palm, for sheer
oddity, may be given to that of the Hag Moth.
On September 18, 1883, this curious brown cat-
erpiller was given me by a friend, taken from a
cherry- or apple-tree. It is brown in color,

FIG. 88.   THE "MONKEY-FACED" MOTH.   HAG MOTH (PHOBETRON
PITHECIUM)

very nearly the shade of an almond meat.   It
is rough in appearance and most singular in
form.   It was a study to find at which end of
him was the head, for, like that of the "saddle
worm" (*Empretia stimulea*) and that of the

Limacodes scapha, to which genus he also belongs, his head is out of sight, under the first ring. There are three singular appendages, flanges or fins they might be called, on each side, as shown in the sketch of him above. The appearance of this "flange" is much like one of the points of a star-fish, even upon close examination, and there is the same little blackish round-dot finish at the base of each, or where it joins the body. Upon examination with the microscope there is to be seen a double row of starry spines (eight on each side) down the top of the back, but so fine (while yet perfect stars) and so exactly the color of the rest of the body as *only* to be noticed with the microscope. Just under the upper surface of the caterpillar (so as not to be seen from looking at it above) there is a row of smaller-sized stars extending around the entire body. Had this been noticed by the entomologist who gave him his name he might have had a prettier one than that chosen from his homeliest feature. For myself I shall call him the Hidden-Star Caterpillar. He is like the Scapha in his movements, gliding along the leaf with a slow, graceful motion; and if disturbed, he puts down his head on to the leaf, bending over and making a low brown bridge of himself. His head is small, amber-color in the centre and dark-brown on either side. He stands on the side or edge of the leaf to eat, bending his head over the leaf so that you cannot see him eat, except by looking on the under side of the leaf. He eats the cherry-leaf readily, and, although his motion seems slow, he goes very soon from one end of the long leaf to the other. It is a difficult thing to see him eat, even when you know by the melting away of the leaf that he is taking a meal, and have also your microscope well adjusted for observation. This is because his mouth is so protected by a fleshy half-hood on each side that you can only see the crescent he cuts growing larger, and his head (wherever it is!) moving along to meet new demands of his cherry-leaf. I have watched the same thing in the "Saddle" caterpillar; his fleshy hood sucks down upon the leaf and hides his mouth entirely.

On the 23d of September, five days after he was given to me, he *began* to throw off his flanges—not all at once, but gradually, one on the 23d and the next on the 26th. On the *edge* of most of these flanges, near the end, there are two small black hairs with little black knobs, like tiny pins stuck in, and from this point they seem to break off most easily. On the 26th, after losing off two flanges, he fastened himself between two leaves preparatory to making his cocoon. On the 28th the hard, almost round, blackish-brown cocoon was finished, in size and shape as given above, and, strangely enough, having its flanges stuck around it by way of ornamentation! I had *read* of this curious habit of the Hag Moth before, and could scarcely believe it possible that he could pick up and attach these appendages to his cocoon after it seemed finished. But "seeing is believing," and there they surely were, and evidently as firmly fixed to the little ball as they had formerly been to his body. I should have seen *how* this was done, at the risk of disturbing him in his cherry-leaf covering, had he not stolen a march on me when I "was busy here and there," and so kept his secret a mystery still.

Watching was now over, and, except to label him in his glass prison, he might be forgotten until the winter was over and gone. Yes, and the *spring* also. On the 5th of June the lid of the little brown house was thrown back a very little, and out stepped this very pretty moth after nearly a ten-months' sleep. He flies quickly from one part of the box to

another, or walks with his funny twinkling
feet rapidly up the glass, a contrast as surprising to the slow-motioned caterpillar as is the
handsome coat he now wears to the rough
brown jacket of his caterpillar days. So far
as I could see, or prove by tempting sweets, I
could not find that he had a tongue. Perhaps
he might have found it himself had he been
free to fly from honey-cup to honey-cup
of real out-door flowers. Yet I think if he
had been very hungry and had any means of
supplying his need, he would not have scorned
the sugared moss that seemed to have no attraction for him. True to his first and second
stages, he is still *brown*, but handsomely *shaded*,
making so much of the *different* shades of his
favorite color as to give him a very handsome
dress. The principal color is almost a seal-
brown (a few shades lighter), and the front
wings marked with dashes or spots of a light
yellow-brown, with a wavy band of the same
color crossing them; a spot of still lighter
brown marks the centre, and the edges of the
wings are a shade of still darker and very rich
reddish-brown. His beauty can only be really
known under the microscope, which brings out
a richness of coloring and beauty of arrangement that redeems the Hag Moth from any

suggestion of homeliness by his most unfortunate and inappropriate name. When he came out from his brown cocoon the skin of the pupa, clear as crystal and perfect in shape throughout, came with him, and lies now by his side in his little box, and the lid of the cocoon flew back so exactly in place that to-day one might look at it carefully and think it a perfect, unbroken chrysalis. Besides the six flanges described, there were also six miniature ones, which scarcely showed beside the longer curved ones, but which, like them, fell off, and were, in part, at least, attached to the little brown cocoon. Just twelve in all, now safe beside the cocoon, the pupa case, and perfect insect. All that is wanting is the triple row of stars, but by one who has once seen them he will be still remembered as the *Hidden-Star Moth.*

XXXVI.

O N the 25th of May, 1878, a bright-col-
ored caterpillar, which I had found on
the smartweed, made a cocoon. It was very
thin, and of an almost glass-like material, except that the *top* was ornamented by fifteen or
twenty of the little knobs of the pink blossoms
on which it had been feeding. The cocoon
was boat-shaped, and the chrysalis inside was
a rich, shiny light brown, tapering from the
head to a very pointed end.

The moth came out on June 14th, after
about three weeks from his change to a chrysalis. He is very delicate in coloring, without
a trace of the rich hues he wore as a caterpillar. The upper wings are a bright silvery
white, dotted slightly with dark gray, the edges
rounded, and delicately finished with narrow
white fringe, just above which is a row of tiny
black dots, each at the end of a crimped line

or fold.   The under wings are silvery-white in
fluted folds, with no color except a row of very
minute black dots above their beautiful silver-
fringed finish.   The body is ringed with alternate silvery-white and brownish-black dashes

FIG. 89.   THE SMARTWEED CATERPILLAR.

on a white ground.   The antennæ are long,
not feathered, and slightly curled at the end.
The joints of the feet next the body are broad-
ly feathered, but slender to the foot itself—
looking like a child's arm in a short puffed
sleeve.

The caterpillar of this moth, as I have said, is very gayly colored. The head is jet black, with a few yellowish-white bristly hairs falling forward from it and from the first (black) ring. The second ring and all the rest but the last two have on them six reddish-brown little knobs, with yellow radiating spines, with two short white lines below each circle of spines. The first ring has two short white lines, the first one of the two interrupted or broken in the middle ⁻ ⁻ and the second whole ; on the second ring this is exactly reversed, the first line whole and the second broken ⁻⁻, and the third is the same ; all the rest are broken and more irregular. On the last two rings there is only a white line. A scalloped line of bright gamboge yellow runs down each side of the caterpillar, and little dashes of yellow here and there over another line of white. The reddish-brown knobs are on a velvety-black ground, giving the whole a very rich appearance. The under part is a dark-brown. Sometimes the cocoon is like thin white silk and almost transparent. The nearest description answering to what I have only known as the "Smartweed Caterpillar," is the *Apatela oblinita*, or, as it is sometimes called, " The Smeared Dagger."

Both the caterpillar and moth very nearly answer to this description, the greatest difference being the transformation, as one writer gives it, "occurring in the ground." I have often raised the moth from the caterpillar and never seen any variation in the form of its cocoon, until a most singular experience last September, which has given me a *rare* "specimen" indeed. A caterpillar from the Smartweed, to all appearance the same as above described, was secured and placed in a glass box. To my surprise, when the time came for his change, instead of spinning his thin cocoon he went up the side of the glass and fastened himself in exact imitation of the Asterias or Turnus butterfly, by a slender thread, with no hint of a cocoon! Here he stayed, not throwing off his skin, but keeping his position so long that I was led to examine carefully into his case. What was my surprise to find that he had changed into a white substance closely resembling chalk. The *head* retained its natural color, being even brighter than before. All the rest was hard and white throughout. Where had the brilliant colors gone? He now sleeps in a gilt box on a bed of pink cotton, a curiosity as well as a *lesson*, for what words could so plainly emphasize the truth that in giving up our own natural way of living in the attempt to imitate others, we shall neither become like them nor keep our own identity, and only remain fit specimens for lovers of the grotesque to place in their cabinet of curiosities.

XXXVII.

THE GREAT LEOPARD MOTH.

ON the 4th of October, 1884, I received from a friend in Orlando, Florida, a handsome, quite large moth, and a large number of eggs which were laid in the box after its capture in the pine woods. The moth was white and covered with black rings and ovals. Its body on the upper part was yellow, with rich and very dark bronze-blue spots on the back and sides, while the under part was white, with black dots to match the upper wings. The hinder wings were white, with a few irregular black spots on the border. The moth was about two inches across. It was left a day or two in the box before sending, and my friend, on looking at it before mailing, wrote : " As the 'white owl' in the box was so still I said, ' If he *is* dead I 'll send him,' and when I looked, what a sight met my eyes! Hundreds of turquoises in layers—in tiers—

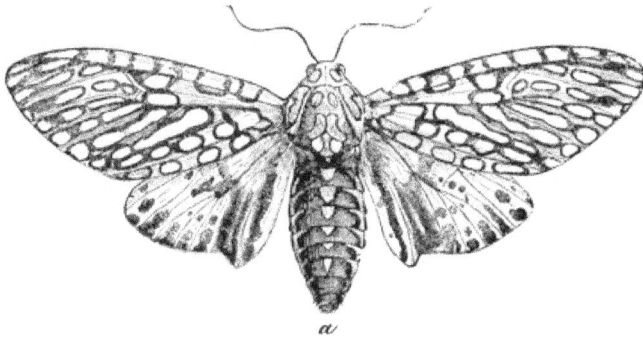

FIG. 90.   THE GREAT LEOPARD MOTH.   [SCRIBONIA].
a, FEMALE ; *b*, MALE.

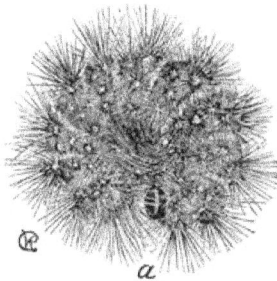

FIG. 91.   CATERPILLAR OF LEOPARD MOTH (CURLED UP LIKE A
CHESTNUT BURR).

and now you can have plenty of white owls.
But what shall they be fed with ?   As they
were found in the pine woods, possibly on the
Blackjack oak leaves."   A part of the "turquoises" were sent to Professor Lintner, who
kindly identified them, and gave as their food
plants, plantain, wild sunflower, and willow,
with one or two other varieties of plants.   He
pronounced them eggs of "The Great Leopard
Moth," Scribonia, which is "the largest of the
American Arctians."   Two days after receiv-
ing them (on the 6th of October) the little
caterpillars came out in hosts, and readily ate
plantain leaves, making fine lace-work of the
leaf, eating only the parenchyma.   At first they
were about $\frac{1}{12}$ of an inch long, amber colored
alternating with dark brown.   A brown head
and one brown ring, next ;   then two clear
amber rings, and three brown with one amber
at the end.   Long black and white hairs
(about evenly divided) were scattered sparsely
over the body.   They would not eat oak or
any other leaves with which I tried them, except mallows, which they ate as readily as
plantain.   On the 27th of October they were
about an inch in length, growing very slowly,
and changing to a rust-red color after the first
month.

On November 3d I gathered a quantity of
plaintain, now difficult to find, and made a
"silo" by packing it closely in a glass jar,
hoping it would last them until they "spun
up." It was a vain hope. The ice and sleet
and snow of late November covered all the
"green things growing," and still the Scribonias lived and still craved food. On November 25th I succeeded in getting enough mallows
and catnip to last a few days. But soon this
was gone and more snow came and still they
lived on! They had had six moults, and
where was it all to end? A supply of spinach
from the grocers, while it was to be had, took
them into December. Now and then in some
shaded nook a little mallows could be found,
even a small "basket-full" is recorded for the
8th of December. On December 9th I "made
over a dozen little cornucopias and placed in
their boxes to entice them to go into chrysalids!" But caterpillars will not be "hurried"
any more than chrysalids, and still they
ate on! Christmas came and went, deep
snows followed, the old year went out and
the new came in, and *still* the Scribonias
lived on. I became discouraged and had
about given them up to *time and fate*
when a card from Professor Lintner, on Janu-

14

ary 10th, announced the arrival of two fine
Great Leopard Moths.   Again I attempted to
satisfy them with cabbage leaf, and as a last
resort (" necessity being the mother of inven-
tion ") they were fed on apples, sliced so thin
as to make them think it was *leaves* with which
they were supplied.   On the 29th of January
the first Scribonia chrysalis was made.   On
the 30th there were seven chrysalids ; on the
31st, eleven.   On the 11th of February I
watched one of the caterpillars change into
the chrysalis.   It took a little over an hour
before the heavy brown coat, with its crimson
bands, was thrown off, and the plain brown
chrysalis was still.

When these caterpillars had reached their
last moult I was surprised to find that I had
once had two specimens of the same, found in
Pennsylvania (Easton).   *They* lived a long
time, but finally seemed so stupid and still
they were given up as useless, and thrown
away.   No doubt had " patience " with them
had its " perfect work " I should much sooner
have known the beautiful Great Leopard
Moth.   They were very similar to the common brown and black caterpillar of the Arctia
Isabella or Isabella tiger-moth, but about
twice the size, and on close looking showing a

bright crimson line marking each ring. Harris says of it : " It has been confidently reported to me that the Great Leopard Moth has been seen in Brookline, but it must be very rare here for I have never heard of its being taken in any part of New England. Specimens of this fine insect would be a very acceptable addition to any collection of such objects." I thought I fully understood that last remark ! I understood it better on *March 4th*, when after all the watching and waiting two fine specimens of the Great Leopard Moth stepped from their chrysalids, and were at once named " Cleveland " and " Hendricks," in spite of my politics, in honor of the two successful candidates, who after an equally long struggle were inaugurated on that day ! After this it grew to be no surprise as one after another left their brown cases until all had made their exit, and "specimens" were at a discount ; although the beauty of the moths paid, after all, for the very leisurely way in which they chose to give this "very acceptable addition " to our collection.

FIG. 92. THE EUDAMUS TITYRUS.

XXXVIII.

A BUTTERFLY CHASE.

ON the 24th of August, 1881, a very singular caterpillar was given me. He was of a pale yellow-green color, with a large red head, on each side of which were two round bright-yellow spots, giving him the appearance of looking at you with very big eyes. The spiracles were small, and black, and the feet orange color. His habits were as peculiar as his looks. He kept closely to the under side of the wistaria leaf on which he fed, although this caterpillar likes the wild bean equally well. He even fastened himself slightly to the leaf by spinning a few threads, to secure him more effectually from prying eyes. But as he ate of the leaf, in nearly circular holes (from near its centre), I could watch the movements of his head from above, yet could not see him actually eating. I watched in vain for this, and came near starving him, by deciding that he did not want food, until I learned his secret, which was that he ate *only* in the night. I kept him until August 31st, when, in some mysterious manner, he slipped the moorings to his leaf, and managed also to get out of his glass prison, and I saw *him* no more. But the *first* lesson was secure. In September (5th), 1887, during a woods-ramble, I found another caterpillar of this kind upon a wild bean (the *Wistaria frutescens*). His red head, with the large yellow spots like eyes, quickly led me to know him. His neck seems set in his head, like a cork in a bottle, only that it turns easily, and reminds one of the neck of a toy needle-box bird. Under the microscope the head is rough, like the rind of a cantelope. He, too, ate only at night. A few hours after I had secured him, on looking at him his head seemed a clear amber- (glass-like) yellow, but on close examination it proved to be only the *old shell of the head*, not detached, but pushed forward, and soon it fell upon the bottom of the box, the new head looking brighter than ever, and very soon the old coat followed, and all was fresh and new. *So* far he got, but failed to make a chrysalis! In May, 1888, four chrysalids were given me by a friend. What they were was not known. They were a dull yellowish-brown in color, very full-bodied, the four or five rings at the end having a screw-like appearance, and looking on the entire surface, with its fine crinkles, like "crackle-ware." On the 26th of May I heard a little rattling noise from one of the four chrysalids, and soon two small black eyes appeared at an opening on the back, just below the head. It was an Ichneumon fly, and the tough chrysalis gave him work enough to pay for his robbery. He thrust out a pair of antennæ, and unrolled an amber tongue, and took sweetened water from his chrysalis-case prison, much to my amusement. He did not succeed in freeing himself until the next day, when I broke a small bit of the case, when he walked out, and about the box, a russet-brown fly, with a black head, smoke-colored wings, black antennæ, with one bar of honey-yellow across them. The under body was yellow, two light lemon-yellow spots on the thorax. The thighs were russet-yellow. He was about the size of an ordinary wasp, and a Beau Brummell for polishing and pluming. What the other three chrysalids held was still a mystery. On June 3d, a week after this Ichneumon appeared,

the second chrysalis opened. I had been out
to drive that afternoon, and as we came near
a wild plum-tree we saw a very pretty butterfly darting among its flowers, and then across
the road like a flash, and back again. The
carriage was stopped, and I eagerly watched
the efforts of one more successful in butterfly
capture by hand than any one I know to
secure him, but in vain. Your hand was upon
him just as he was on the other side of the
road ! Securing some branches of the sweet
blossoms, I was arranging them in a vase on
my return, on the table where the three
chrysalids were lying, when the second chrysalis opened, and out came the very butterfly
we had failed to secure in our afternoon's
chase ! The wild plum-blossoms were ready
for him ! The next one gave also a fine
butterfly ; and one never opened. And this
was the way I learned the whole history of
the beautiful *Eudamus Tityrus*.* I have the
two specimens and the Ichneumon now before me, the only ones I have ever secured.
The butterfly is so swift in motion, with such

* The *Eudamus Tityrus*, one of the *Hesperians*, or *Skippers*, is the
largest of the butterflies in that large group, which seems almost
like a connecting link between moths and butterflies, their chrysalids being shaped like those of moths (conical in form), while the
antennæ are hooked at the end, as are those of the sphinges.

a darting, zigzag dash that it is next to impossible to catch him in flight. His wings are a rich velvety-brown, with a golden-edged, interrupted, honey-yellow band across the middle of the upper pair ; and lighter honey-yellow spots (almost in small squares) are found near their tip. The hinder wings have a very short, rounded tail, and a broad band of silver glistens on the middle part of their under side. The antennæ are turned back at the end like a hook. The body is a rich purplish-brown, and the wings are finished with a shaded brown fringe.

FIG. 93.

CATERPILLAR OF EUDAMUS
TITYRUS.

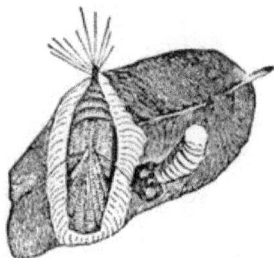

FIG. 94.
CHRYSALIS OF EUDAMUS.

FIG. 95. WHITE-LINED MORNING SPHINX.

XXXIX.

TWO SIDES TO A SHIELD. THE WHITE-LINED
MORNING SPHINX. [DEILEPHILA LINEATA.]

THE first caterpillar of the White-Lined
Morning Sphinx, which I obtained on
September 3, 1881, gave me my first lesson in
the great variation there often is in larvæ of
the same moth. The one I had secured had
three stripes down the back, made up of
shaded, bead-like spots, strung on a line of
pink, with a black line each side of the pink
one. He was very dark, had a sharp horn on
the end of his body, and spiracles black, edged
with yellow. His photograph is given on p. 221.

I had seen this caterpillar in Professor Riley's
"Third Annual Missouri Report," as I thought,
and turned to page 141 to compare my caterpillar with the figure given there. To my
surprise and disappointment, at a hasty
glance I decided that they could not be the
same. In too much haste to read the text,
I was just about closing the pamphlet when
my eye caught the words: "The most common form of this larva is given at Fig. 61."
The next thought, "perhaps he gives *another*
form," led me to turn the leaf, and lo! there
was my caterpillar, without a shadow of doubt,
in Fig. 62. Then every word of his beautiful
description was carefully read. He says:
"Few persons are aware what this beautiful
moth looks like or what it feeds upon, in the
caterpillar state. . . . The very great di-
versity of form and habits to be found
amongst the larvæ of our butterflies and
moths has much to do with the interest which
attaches to the study of these masked forms.
I am moved to admiration and wonder as
thoroughly to-day as in early boyhood every
time I contemplate that within each of these
varied and fantastic caterpillars . . . is locked
up the future butterfly or moth, which is
destined, fairy-like, to ride the air on its

gauzy wings, so totally unlike its former self. Verily the metamorphoses of the lower animals must prove a never-failing source of joy and felicity to those who have learned to open the pages of the great Book of Nature." *Joy and felicity*, the very words for me, and every new specimen served to emphasize them. He then adds: "The White-Lined Morning Sphinx presents one of the most striking cases of larval variation," asserting that from these very differently marked larvæ the moths reared from them "show no differences whatever."

To *prove* this, which was not doubted, was a pleasant task. My specimen went into the box of earth prepared for him on September 7th, and on rolling back the earth from him September 13th I found him a fine chrysalis.

On the 22d of September I secured another caterpillar of the same kind, and afterward three of the other kind, described in Fig. 61. As these made their several changes to the imago, I found the moths were all alike. On the 24th of May the one I specially watched came out from his chrysalis. He was (very unlike the Polyphemus) less than a minute in coming out! I should not have seen it, but happening to look at the chrysalis at that moment, I heard a slight crackling noise and saw a tremor

or shiver go over the upper part of the long
yellowish-brown and pointed chrysalis (where
was the head and thorax of the moth).   Then
the front piece lifted and gave way, and out
stepped a thing of marvellous beauty, which
had been shut up in darkness and silence in a
homely casket since the previous September!
The front wings are a beautiful shade of olive-
green, with a central line across of interrupted black and white, with a hint of rose or
watermelon-heart color.   Each side of this,
separated by a band of olive about the six-
teenth of an inch wide, is a larger, spotted,
interrupted line of black and white spots in
almost squares, the black a little larger and
like velvet, and each side of this is a tint of
rose.   The beautiful rose-colored under wings
unfolded slowly (as did the upper ones in
getting their full expansion).   The legs are
spined and of a delicate mouse color, end-
ing in a minute black claw.   The spines on
the first pair are very delicate and hair like,
and are black.   There are two spines of
unequal length, almost at right angles, and
mouse-color, like the legs, on each side of the
other two pairs.

The head is brown-olive in color, dashed
with white in stripes running downwards, with

a pinkish-white border finish. The side-pro-
tectors of the coiled tongue are so prominent
as to look like a front part of the head. The
thorax is olive-green with white-dashed lines—
one in the centre, then two below it, and a
double V-shaped line on either side.
The antennæ are black, bordered
on the entire outer edge with white.
The side pieces to the tongue are
also white-edged, running back and
forming an unbroken line with the
head markings. The eye is deep
set, the pupil dark
and perfectly round
in the centre of
the mouse-colored
wheel-like eye. The
tongue shows a little
dark wheel between
the side pieces. The
body has a middle
line, lengthwise of

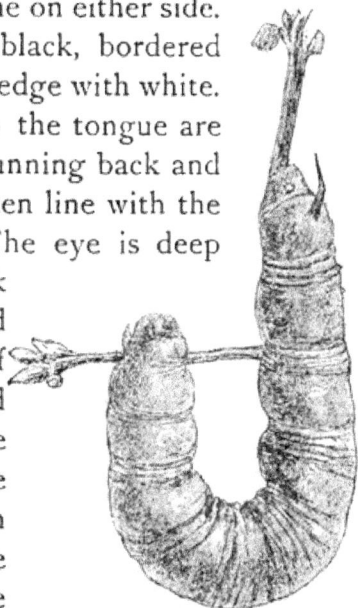

FIG. 96.

interrupted, short, black lines on a soft, mouse-
colored ground, and either side of this central
line is a row of round black dots, to the end
which is very pointed and finished with a
pencil of rich brown hairs.

The antennæ in shape look like small bean

pods. They are crossed by regular lines. The centre a chocolate color, with a finely-toothed white edge. Both pairs of wings are elegantly fringed with white. He is swift in motion and when disturbed makes his wings twinkle, as when hovering over a flower. His generic name signifies "Evening Friend," and he is seen to the best advantage when flitting with perfect freedom, like a humming-bird, from flower to flower, sipping sweets with his long tongue and making the most of his new-found higher life.

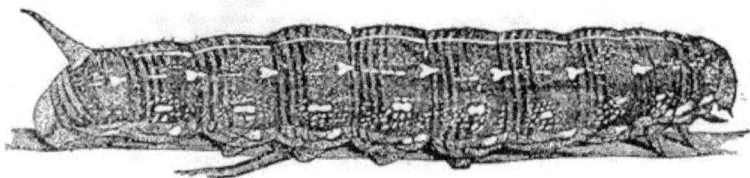

FIG. 97.

FIG. 98.

XL.

THE "DECEPTIVE" MOTH.

MORE like a very modest, trustworthy Quaker looks the richly mottled silvery-gray moth *Apatela Americana* which stole in upon me on Sabbath, April 20, 1890, from its plain gray parchment-like cocoon;

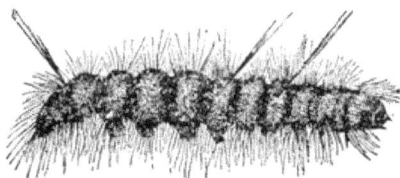

FIG. 99.

so stiff and hard an one, that in making his exit the whole of the rich brown chrysalis inside it was thrown out into the box beside the clinging moth. Still, demure, quiet, why should it be named "Apatela—*deceptive*"?

Because some entomologists are "strict to mark iniquity," and allow the *moth* to bear the blame of the *first* stage of its existence. For the *Apatela Americano* is an "*Owlet*" moth, one of the true "Noctuæ" tribe, and yet its caterpillar is so like that of the Arctians as to deceive one versed in entomology, and to lead him to expect either a "great bear" or some other Arctian in the imago. I confess to having been "deceived" by the *very* caterpillar which produced this moth, having bought him of a little boy on August 14, 1889, thinking at the time, "I have had *enough Arctians*, but this one is so bright in *color*, that I will try one more." He was of a fine sulphur-yellow color, and was placed by some light straw-colored Arctians in a box to which little special attention was paid. True he had some long black pencils—two of them on the fourth ring, two on the sixth, and one on the next to the last. But some of the Arctians looked much like him. However, when he began "spinning up" on the 16th, two days later, I soon saw a difference in his *cocoon* and that of the Arctians. Those, instead of being rough and parchment-like and nondescript in shape, are smooth (hairy) and oval—looking as if *evenly* sheared to one exact length throughout.

How long before he changed into a chrysalis inside his rough hammock I could not peer within to see ; but when, after lying perfectly still from August to April, he came out of his hiding place I looked *beyond* the chrysalis to the farther end of the cocoon inside to see if the sulphur-yellow color was visible in the caterpillar robe, which is always folded like a napkin, "in a place by itself,"—and lo ! there it was—the jetty black head and yellow coat, making assurance doubly sure. He now stands quietly under the glass beside me, having satisfied himself with a full meal of sugared water, which I watched him take with his broad flattened, amber-colored tongue. His antennæ are long and slender, have more of a *twisted* than ringed appearance, and are inserted in a little round socket, just above his large dark seal-brown eyes. His body is deeply ringed, and of the same soft brown hue as the wings. These are handsomely shaded, and their borders elegantly fringed, with white intermingled with lines of black. The first pair of legs are ringed with white and black lines and puffed at the top like an old fashioned "muttonleg" sleeve. The second pair are also ringed but not puffed, and have one small spine. The last pair are plain—neither puffed nor spined.

15

He is so gentle that the deception of his caterpillar state may be forgiven, and he is no doubt as welcome among his "Owlet" companions as though he started in life in a livery especially his own.

XLI.

THE ROYAL WALNUT MOTH.

" WITH patience wait for it," were the first words which came into my mind as, in the night of May 5, 1889, a slight tapping noise attracted my attention. On looking in the direction of the sound, I found the stranger, who first knocked and then entered into the world without waiting for a friendly "Come in," was no other than a beautiful Royal Walnut Moth (*Ceratocampa regalis*). "With patience," because, for eleven years, I had waited in vain for the perfect imago of this rare and beautiful moth.

The first caterpillar of this species was given me on August 30, 1878. After going through his moultings successfully, and forming at length a perfect chrysalis, he failed to appear, and remained in his casket without power to reveal what "might have been."

Again and again other specimens were se-

cured, and carefully watched through different
changes, but all died before the perfect insects
appeared.   On September 6, 1888, a fine
specimen was given me by a friend ; and this,
after more than eight months' delay, is now
the beautiful *Ceratocampa* before me.   Looking back at a record made on September 8th
of that year, I find this entry : "Watching
my Royal Walnut. He eats silently and
rapidly, the walnut-leaf melting away in front
of him.   He clasps the leaf with his first pair
of russet-colored feet, and eats downward, so
that his head bends toward the ground.   The
last two pairs of his long-spined horns lie back
gracefully.   The first short pair stand forward
like ears.   The second pair lie across the
third, now, as he eats.   He eats so as to leave
a crescent in the leaf.   The long narrow point
of the leaf shakes like an aspen as he eats, until he cuts it off and drops it.   There are three
round black dots on each of the two last pairs
of horns on the little yellow part which is next
to the head.   The three pairs of horns are
tipped with black.   There are two pairs of
horns on the second and third segments.   The
long point of the walnut-leaf, which he could
not eat (being unable to hold it, because it is
so delicate), he took with his fore feet, and

lifted it gently out of the way, and then began in a new place."

For the next day the entry is: "The Royal Walnut keeps very still. Has lain for a half hour in the same position—head bent down, so that the first pair of horns rest on the floor of his prison." Upon September 10th, "I gave my Royal Walnut his last meal." At noon he was walking slowly on the earth with which a large box had been filled for him. After an after-dinner nap, I again went to his box. The untasted spray of walnut-leaves lay unwithered on the surface, but no trace of the caterpillar was to be seen. Not a movement of a grain of earth above him. He had buried himself.

After a month had passed, curiosity overcame prudence, and the earth was shaken back to see if a perfect chrysalis was below. "There he lay in his imperfect, half-rounded bed—made by moistening the earth about him,—and as still as if dead."

The chrysalids of many moths will be seen to show frequent signs of life ; but the stillest of all still things is the chrysalis of the Royal Walnut. You may watch it for days and weeks, or even watch its shadow, and you will see no slightest movement. The smooth, plump, black head, with its two slanting breathing-holes, is as still as a rock, and its rings (with the two queer flat little humps on the front one) are as still as the head.   Again and again you say : " If there is any life in it, how can it keep so still ? "   Then you satisfy yourself by stroking it very gently, with the faintest touch of your finger, along the side, and lo, a little cringe, showing the slightest shrinking from the touch.   That is all.   Again it is as still as a rock.   After long watching, another stroke, and another almost imperceptible cringe.   It bides its time.   So must you.

The eggs of the Royal Walnut closely resemble the Malaga grape in shape and color. They are clear (unlike those of the Luna and Polyphemus moths)—so clear that the larvæ can be seen through the delicate amber shells long before they are broken for exit.   At first the caterpillar is nearly black.   It changes in appearance, however, with each moulting, at one time being pale-green, again almost a chocolate, and finally a deep dark-green, with pale bands of blue.   The ten spined horns with which it is armed give it a menacing and formidable appearance, but it is at all times harmless.   It is curious to note the different

expressions used by those who look at it.
" Horrible creature ! " one exclaims. " It is
almost beautiful—so richly shaded," says another. One writer says of this caterpillar : " It
is handsomer than the beautiful moth it pro-
duces." But, although it has rich colors, curiously shaded, I should say it took some nerve
to see the beauty, as the form is certainly un-
attractive. That from so formidable a creature
such an exquisite moth should be produced

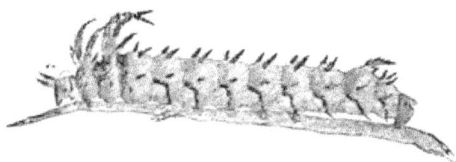

FIG. 100. YOUNG CATERPILLAR.

seems little less than a miracle. In color the
moth is entirely different from the caterpillar.
Its fore wings are of a grayish-olive color,
veined with lines of a peculiar shade of red—
best described, I should say, as nacarat red.
The hinder wings are red, with yellow spots
of irregular form in front, and olive-colored
spots behind, between the veins. The thorax
is yellow, bordered with red. The antennæ,
or " feelers," are amber-colored, and in the
female specimen which I have, appear to be
ringed, when viewed by a microscope.

The moth is gentle and quiet. It takes no notice of offered sweets, and shows no sign of possessing a tongue. For a short time it gives its silent beauty to please, makes provision for other silently beautiful moths (one hundred and twelve eggs were laid by this one), and dies.

The most touching thing in the life of the Royal Walnut is its self-burial. This was carefully watched and timed in one specimen (which, however, failed to develop an *imago*).

FIG. 101.   FULL-GROWN CATERPILLAR.

I will close this sketch by a quotation from a record, kept at the time, of two Royal Walnut caterpillars, one of which thus buried itself : "On the 30th of August, 1882, I was fortunate enough to find two specimens of this caterpillar on a large walnut-tree. They were of a mulberry-brown color (probably being in their second stage), with heads of glassy brilliancy ; brown feet, striped with black ; and light, diagonal side stripes separating the spiracles or

breathing pores. Both were watched through
their last moultings, and one of them changed
into a chrysalis on the surface of the earth in
his box. He had taken no food for a week
previous, and the opportunity of watching him
make the chrysalis was unique and full of interest. He lay upon his back with feet upper-
most, and the head of the chrysalis appeared
earliest. It was large, and of a delicate pea-
green at first. The small, old brown head of
the caterpillar is now gliding down very slowly
on the top of the newly-formed chrysalis, as it
lies on the spined horns below, and looks so

FIG. 102. CHRYSALIS OF ROYAL WALNUT MOTH.

meek and helpless as it is pushed down by the
retreating skin. The sides of the chrysalis, as
they appear, are tinted with pale red. The
spiracles are oval and brown-bordered ; the
antennæ stand out clear amber. Looking with
my microscope, I can see the immature parts
of the moth's head arranging themselves ; the
part where the head is, and inner part of the

vest, not yet being closed. If this space closes over (as it seems to be closed in a perfect chrysalis), it will be very strange to see how it is done. The other Royal caterpillar is eating his leaves contentedly on the walnut branches above him (he is on a spray growing from a bottle of water in his prison), in blissful ignorance of his own coming change."

This chrysalis was not as perfect as those formed underground. That of the second, which buried itself, is the one shown in the picture. The record of its change is under date of September 13th :

" I watched my Royal Walnut bury himself. About half-past eleven A.M., I saw he had done eating, and was very restless, so I put him on a box of earth. It was a touching sight to see him take charge of his own funeral. Slowly he walked around, surveying the ground ; and then, at one corner, chose his lot, and began going down, very slowly, head first, and a little way at a time. He would raise up the back part of his body, nearly vertically, every little while. This earth was fine and mellow, and I thought how difficult it must be for him to go down into the hard ground under the walnut-tree. Nature is wonderful in her workings : Why do the Polyphe-

FIG. 103. THE ROYAL WALNUT MOTH.

By permission, from Flint's edition of "Harris on Insects Injurious to Vegetation."

235

mus, Luna, Cecropia, and Prometheus make
cocoons, while the moths of the Grape, Tomato, and Walnut bury themselves in the
ground? Why does one never change its
own way, and try another's plan—some pre-
ferring a tomb, and others a burial? Ten
minutes past twelve,—forty minutes in all,—
and the last speck of green and brown had
disappeared. By close watching, with a magnifying glass, I learned a new and wonderful
thing. I saw plainly the reason he did not go
down faster. He was making a smooth, soft
tunnel for himself! He threw from his mouth
quantities of water or mucilage, and thus soft-
ened and worked the earth, until the whole
tunnel was really plastered, and then, by a
succession of strong upheavals, he threw the
dry earth over the back part of himself (rather
than draw that in), until he was hidden from
sight. The earth above him trembled and
moved for several hours after, as if he was
still at work in his burial-place below."

The oval earth-casket which this caterpillar
made was much more complete than the one
which held the chrysalis of my Royal Walnut
Moth. It was probably partly from the gentle
breaking of this to get the chrysalis, and from
the jarring in taking its likeness given in the

picture, which prevented the appearance of the perfect insect. One who witnesses the wonderful transformation from the creeping, ungainly worm to the exquisitely dainty moth, winged and fitted for a higher life, is reminded of the words of Scripture: "It doth not yet appear what we shall be."

www.ingramcontent.com/pod-product-compliance
Lightning Source LLC
Chambersburg PA
CBHW021516210326
41599CB00012B/1283